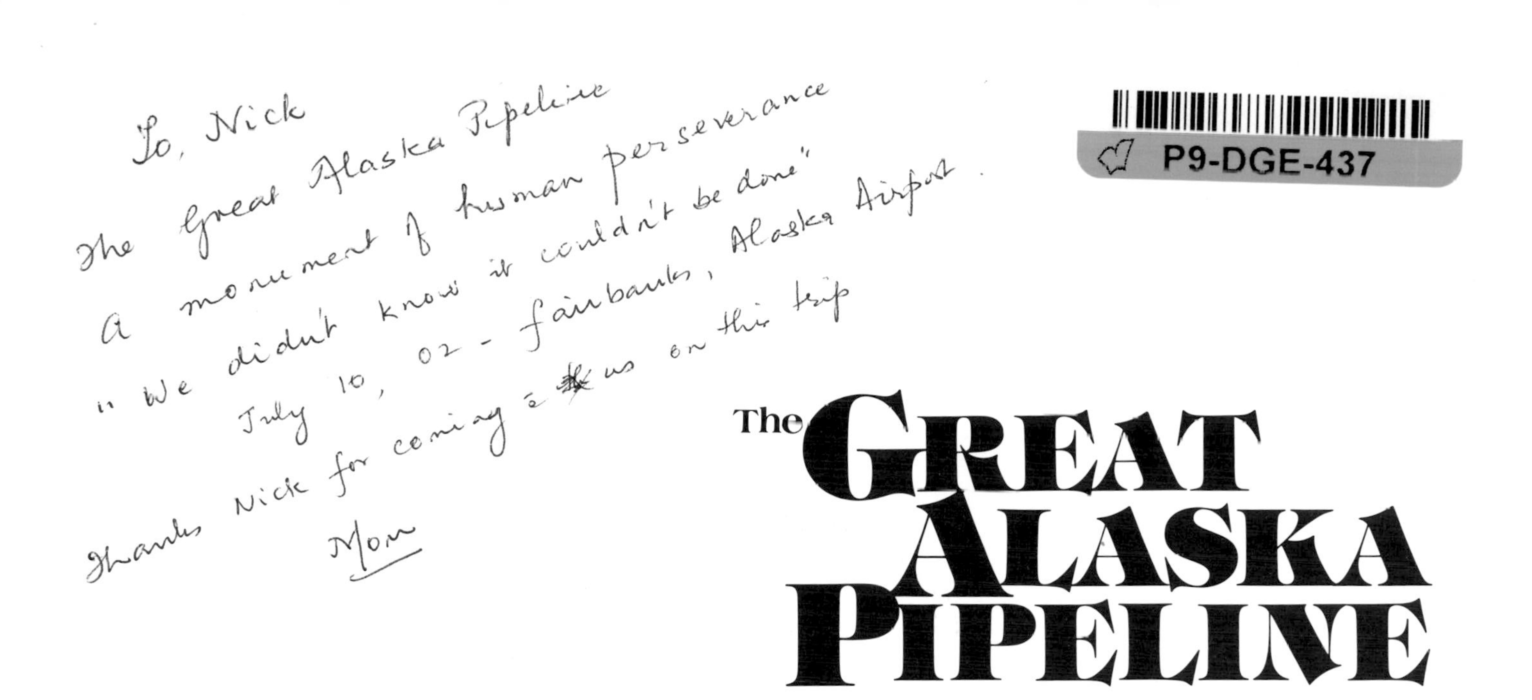

The Great Alaska Pipeline

The Great Alaska Pipeline

BY STAN COHEN

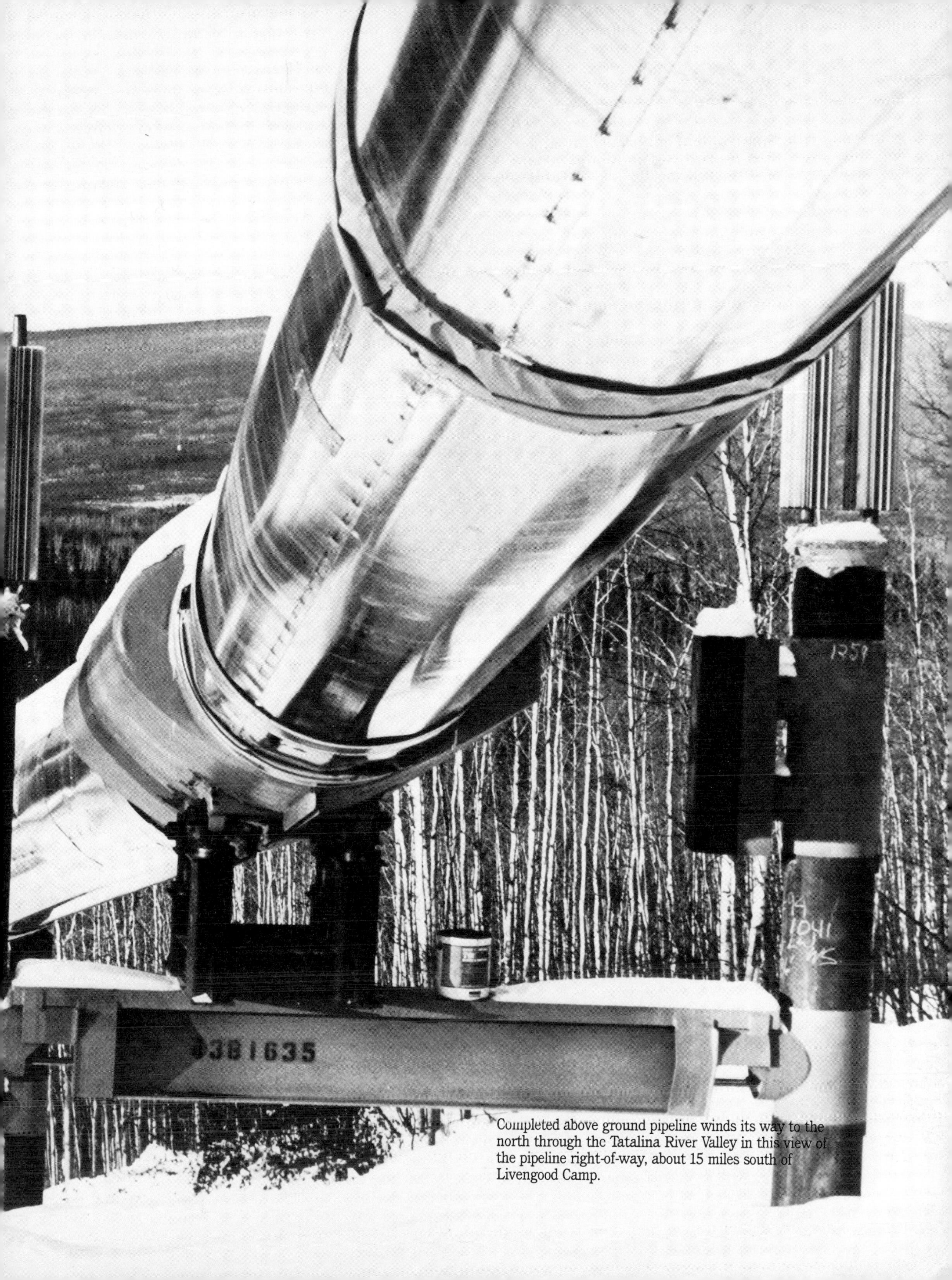

Completed above ground pipeline winds its way to the north through the Tatalina River Valley in this view of the pipeline right-of-way, about 15 miles south of Livengood Camp.

LIBRARY OF CONGRESS
CATALOG CARD NO. 88-60470

ISBN 0-933126-71-9

First Printing: April 1988
Second Printing: April 1990
Third Printing: March 1993
Fourth Printing: March 1996
Fifth Printing: March 1997
Sixth Printing: April 1999
Seventh Printing: March 2001

Typography Arrow Graphics, Missoula, Montana
Cover Art Mary Beth Percival, Missoula, Montana

PRINTED IN CANADA

ALL PHOTOS COURTESY OF
ALYESKA PIPELINE SERVICE COMPANY UNLESS NOTED OTHERWISE.

About the Author

STAN COHEN, a native of West Virginia, is a graduate geologist from West Virginia University. He worked in the oil fields in his native state and for the U.S. Forest Service in Montana and Alaska. He now resides in Missoula, Montana, with his wife, Anne. For the past 23 years he has run his own publishing company. He has written 65 pictorial books and published over 200 titles. Over 20 of these books are on some phase of Alaska's history. For a complete catalog e-mail at *phpc@montana.com* or write to address below.

PICTORIAL HISTORIES PUBLISHING CO., INC.
713 South Third Street West, Missoula, MT 59801

Introduction

ALASKA HAS ALWAYS BEEN America's superlative. It has the highest mountains, the largest free-flowing rivers, the harshest environment, the most abundant reserves of natural resources, the wildest wilderness, and perhaps the only pioneer spirit left in the culture.

So it's not surprising that Alaska is also the site of some of the most ambitious construction projects in history. The Alaska Highway, the largest such undertaking of World War II except for the Atomic Bomb, connected the region to the rest of the nation. The Alaska Railroad, built in the late 1910s and early 1920s, proved that imagination and engineering savvy could overcome a hostile climate. These enterprises—along with the Canol Pipeline, which funneled oil in the 1940s from the remote Canadian Northwest Territories to Whitehorse in the Yukon—made it logistically possible, and gave engineers the confidence to complete the largest private construction project in the world—the trans Alaska pipeline. Stretching 800 miles from Prudhoe Bay to Valdez through some of the fiercest terrain on the planet, the pipeline is a masterpiece of human perseverance. Eight billion dollars and four years of labor on the part of more than 70,000 people—from pipefitters, welders and equipment operators in the field to the executives in the office—were required to produce this massive achievement.

My attempt here is not intended to be the definitive word on the subject. That book—James P. Roscow's excellent *800 Miles to Valdez, The Building of the Alaska Pipeline* (1977, Prentice-Hall)—has already been written. Rather, I have chosen to produce a visual story using photographs.

I wish to give my heartfelt thanks to the staff of Alyeska Pipeline Service Company's Public Affairs Department in Anchorage, especially John Ratterman and Kim Peterson, who provided most of the photographs and graphics you see here as well as much of the research material. They also reviewed my manuscript for accuracy and offered many useful suggestions for its improvement. My good friend Ken Brovald of Anchorage, who worked in transportation during the Pipeline's construction, provided additional research material. And once again, Mary Beth Percival of Missoula, Montana, put forth a supreme effort to provide a striking cover.

STAN COHEN

Table of Contents

1. From oil seeps to Prudhoe Bay: A short history of oil and gas in Alaska . . . 1
2. How to get the crude from Prudhoe Bay to the Lower 48 . . . 11
3. On again, off agian. Will the pipeline be built? . . . 15
   Building the Pipeline Haul Road . . . 21
   Construction Camps . . . 24
4. "We didn't know it couldn't be done." The construction process . . . 33
   The Yukon River Bridge . . . 34
   Archaeological Findings . . . 39
   Transportation Facilities . . . 40
   Construction Techniques . . . 44
   River Crossings . . . 61
   Welding . . . 68
   Antigun Pass . . . 74
   Thompson Pass . . . 76
   Keystone Canyon . . . 80
   Communications . . . 85
   Animals . . . 87
   People . . . 88
5. Pump Stations . . . 89
6. Pipeline Terminal at Valdez . . . 111
7. The Pipeline Monument . . . 125
8. End of Project . . . 128

Malcolm Alexander's 'Monument to a Monument' at Valdez. Dedicated in 1980, it is a memorial to all the men and women who built the trans-Alaska Pipeline. It features a native, a woman, a welder, a surveyor and an engineer.

1. From oil seeps to Prudhoe Bay: A short history of oil and gas in Alaska

THE EARLY HISTORY of Alaska was shaped by the exploitation of the territory's vast natural resources, mainly gold. It wasn't long after the first placer gold was found at the turn of the century that Alaska had earned for the U.S. Treasury more than the paltry $7.2 million Russia was paid in 1867 for "Seward's Folly."

The first prospectors also found large deposits of nickel, cobalt, silver, lead, molybdenum, zinc, tin, tungsten, iron and chromium, plus non-metallic minerals like asbestos, barite, gypsum, jade, clay, marble, sand and gravel. Huge tracts of virgin timber made lumbering one of the state's early leading industries. And the fishing off Alaska's shores is unsurpassed in the hemisphere. With wealth like this it's no wonder that Alaska has always been close to the hearts of people in the Lower 48.

But it was the discovery of commercial oil deposits, first in the Kenai-Cook Inlet area and later on the Arctic slope, that dramatically upped the stakes and changed the character of the state forever. Alaska is no longer simply a factor in the continued prosperity of the U.S.; it is the key.

America's first underground oil deposits were brought to the surface through Edwin Drake's well at Titusville, Pennsylvania, in 1859. The gigantic Spindletop and East Texas fields were discovered in the early 1900s, and soon oil assumed its number one rank in the American economy that it still holds today.

In Alaska, oil and gas seeps were first detected by the Eskimos near Cape Simpson and on Barter Island, long before the first white explorers moved into the region. The natives cut blocks of oil-soaked tundra from the ground and used them as fuel. In the 1850s Russian colonists reported seeps near the Iniskin Peninsula on the west shore of Cook Inlet. And other seeps were observed on the Alaskan Peninsula and in the Katalla and Yakataga areas on the Gulf. The first claims were staked on this oil as early as 1896—just when gold was discovered in the Yukon Territory and the region flooded with greed-crazed prospectors looking for any way to make a buck.

The first drilling took place in the Iniskin area in 1898—a very early date considering the remoteness of Alaska. Just after the turn of the century more wells were drilled at Katalla, about 50 miles southeast of Cordova—on Cook Inlet—near Nome, and on the Alaska Peninsula. Ernest de K. Leffingwell, a prominent geologist and explorer, spent eight years, from 1906 to 1914, mapping and studying the Arctic area. A U.S. Geological Survey geologist, F. C. Schrader, crossed the Brooks Range in 1901 and made a geological exploration of the Arctic coast. These studies led to more activity. And the 1907 sale of oil to the Copper River and Northwestern Railroad under construction near Katalla ignited a boom reminiscent of the Alaska gold rushes. Almost 10,000 people poured into the area and a refinery was built. In these

Grammer #1 well, Salmon Creek, Iniskin Peninsula.
AMHA

The oil boom town of Katalla. AMHA

Oil well at the Katalla field. AMHA

heady first days as much as 154,000 barrels of oil were produced from a dozen shallow wells drilled by the cable tool method. The burning of the refinery in 1930, however, spelled the end of the Katalla oil industry.

In 1910 all lands in Alaska were withdrawn from oil and gas leasing by Congress. But after it rescinded its decision in 1920 with an oil and gas leasing act, the race was on again. Renewed activity centered on the Chignik district on the Alaska Peninsula, in the Cook Inlet-Susitna Valley and on Admiralty Island. In the 1920s wells were drilled in the Kanatak district on the Alaska Peninsula, the Yakataga district, on the outskirts of Anchorage and near Chickaloon in the Matanuska Valley. In the late 1930s the first two wells using rotary drilling rigs were put in, one on the Iniskin Peninsula and one in the Kanatak district. Neither, however, produced oil in commercial quantities.

When U.S. Navy ships converted to oil power in the early 1900s the government began setting aside oil reserves to insure an adequate supply for the future. The fourth of these depots—Naval Petroleum Reserve #4—was a 37,000-square-mile tract in the Arctic area of

Kanatak on the Alaska Peninsula in the 1920s. Scene of a major early oil boom. Oil company freight yards can be seen at the right. AMHA

Main Street of Kanatak, 1923. AMHA

A Standard Oil Company tractor and trailer at Kanatak, 1924. AMHA

Coastal Drilling Company's oil rig on Swanson River on the Kenai Peninsula, late 1950s.
AMHA, ALASKA RAILROAD COLLECTION

An oil drilling derrick on the Arctic slope, 1951.
AMHA, ALASKA RAILROAD COLLECTION

Shell Oil Company's offshore oil rig in the ice of Cook Inlet, late 1950s.
AMHA, WARD WELLS COLLECTION

An early view of the Prudhoe Bay field before the pipeline was constructed.

northern Alaska established in 1923. During World War II some oil exploration occurred here but nine years and 37 wells later the search ended with only minor success.

The modern era in oil exploration along the Arctic slope began in the 1950s. By this time the geologist had the advantage of modern technology and didn't have to live off the land or spend years in the wilderness like his predecessors. Alaska's first truly commercial oil field wasn't developed until 1957, when the Richfield Oil Corporation began exploiting fields at Swanson River on the Kenai Peninsula. Exploration was extended offshore into Cook Inlet, and by the late 1960s five fields were producing oil and nine producing natural gas in the Kenai-Cook Inlet area. By 1975 Alaska ranked seventh among oil

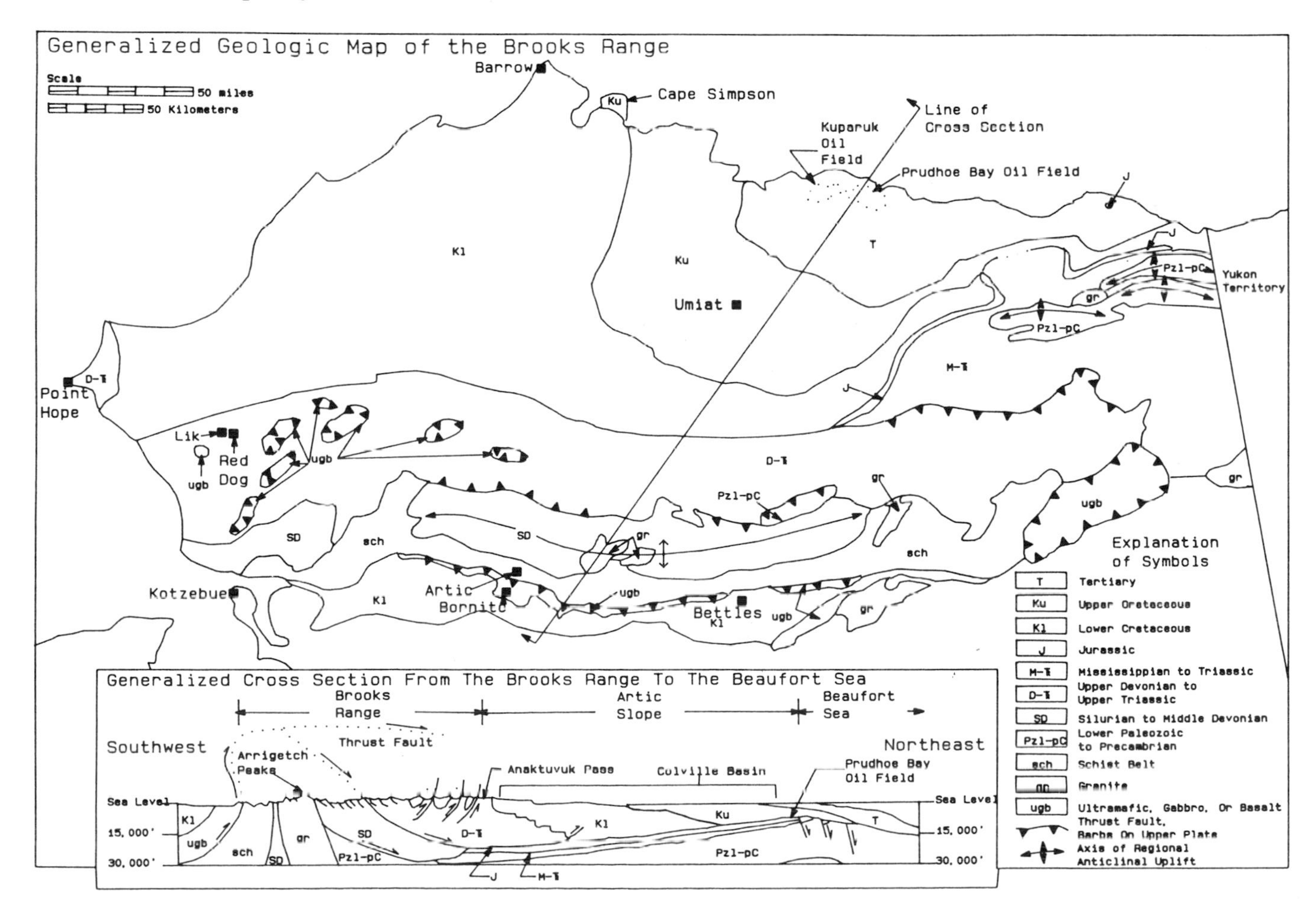

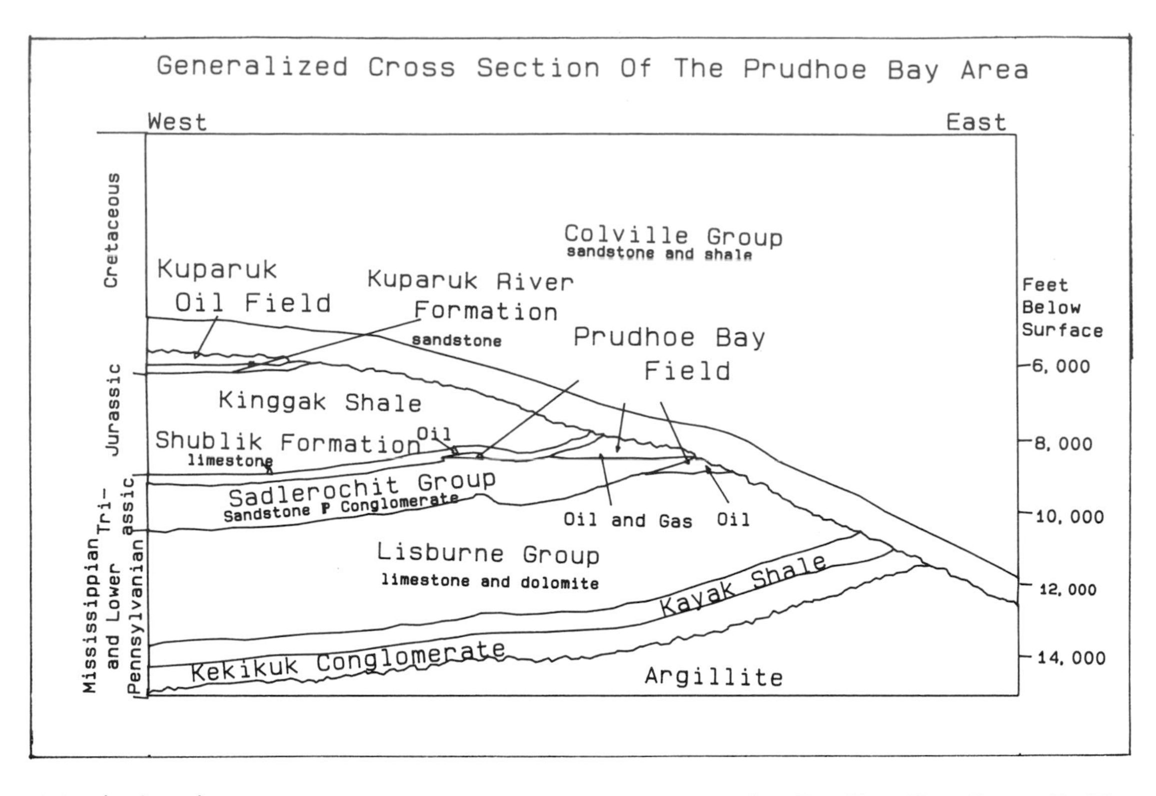

producing states in America.

The 1959 Statehood Act entitled Alaska and its native population to split up millions of acres of what had been federal land. State officials selected a 1.8-million-acre wedge of the Arctic coastal plain between the Naval Petroleum Reserve to the west and the Arctic National Wildlife Refuge to the east on the advice of the state geologist, who saw in his geological reports of the vicinity some similarities to petroleum-producing formations in the Rocky Mountain area.

In December 1964 the State of Alaska offered oil and gas leases in the Colville River delta. Two companies, British Petroleum and Sinclair Oil, obtained large tracts on which two wells were drilled. Seven months later a second lease sale was held with land offerings in the Prudhoe Bay area. At this sale, Richfield Oil, now the Atlantic Richfield Company, and Humble Oil, now the Exxon Company, as partners, picked up more than 71,000 acres on a crest of a subsurface structure. British Petroleum also acquired leases lower down on this structure.

The corporations had their land, but how were they going to get their enormous drilling rigs to these remote sites? The problem was solved by a novel idea. Alaska Airlines leased a military C-130 Hercules cargo plane and the ferrying began, the first use of a military C-130 for civilian use. Many wells were drilled on the North Slope, but they were all dry holes. A 13,500 foot deep well, the Susie #1, was drilled in 1965, 80 miles east of Umiat in the Naval Reserve. It turned out to be a dry hole. The rig was moved overland 60 miles north to a new site, Prudhoe Bay State #1. The well was spudded (started) in the fall of 1966 but because of melting tundra work could only be done during the winter months.

By December 1967, the partners had reached a sandstone/conglomerate formation—the Sadlerochit Formation at 8,202 feet. Testing showed a strong indication of hydrocarbons. The drill went deeper. And finally, it hit oil. By March 1968, 1,300 barrels of oil per day were pouring from State #1. Geologists knew that they'd hit something big, but they didn't know how big.

A second well was drilled into the Sadlerochit Formation and it too tapped great quantities of oil. The partners hired the prominent international consulting firm of DeGolyer and McNaughton to evaluate their find. Their estimate of five to ten billion barrels of recoverable oil reserves in this field astounded the petroleum industry. That figure made this find the biggest in North America, bigger even than the East Texas field. Since then further studies have put the figure closer to ten billion barrels of oil and 26 trillion feet of natural gas. Those numbers represent one-third of America's known reserves of oil 12 percent of its known gas reserves. Another large field was discovered west of the Prudhoe Bay field at Kuparuk. It showed reserves of 1.3 billion barrels of recoverable oil from a shallower formation than its neighbor.

The oil was there. Now the only problem was how to get it home.

Y TO SPEND MILLION ON ALASKA OIL

Plot To Wreck

Army Engineers To Build Special Whittier Tunnel For Oil

Anchorage Daily Times

TOKYO GAS CONTRACT ASSUR

HUGE PLANT SLATED FOR KEN

Pan American Wil

Tests At 1,50

Major Prod

To Join In

Bartlett

Anchorage Daily Times

S BIG DEVELOPMENT OF MATANUSKA

FEDERAL COMMISSION TO MAKE INQUIRY INTO FEASIBILITY OF ESTABLISHING BIG COAL BASE

Anchorage Daily Times

ARCTIC SLOPE WELL STRIKES OI

Minority Leaders Attack Shuffling Of Agency Chiefs

'It Looks Good,' Says Richfield, At Prudhoe Bay

Heavy Fight Is Continuing For Citadel

Anchorage Daily Times

ENATE OK'S PIPELINE

AS AGNEW BREAKS TIE

Jackson Bill Goes With Added Punch

Senate Nod Came Today In 5 Steps

Panel Okays Bill

Anchorage Daily Times

SOHIO HITS RECORD WELL

Viet Troops Smash

; Now At Dong Ha

Discovery Is Largest On Slope

'Localizing' Move Dies

AD HOC'S LEADERS QUIET ON SUPPORT

Havelock Sees Misunderstanding On Residency

JAPANESE LEADERS WILL VISIT ALASKA

New Air 'Triangle' To Boost Tourism

Teachers Accept 2-Year Contract

Assembly Okays $56.8 Million For Schools

Weather Forecast

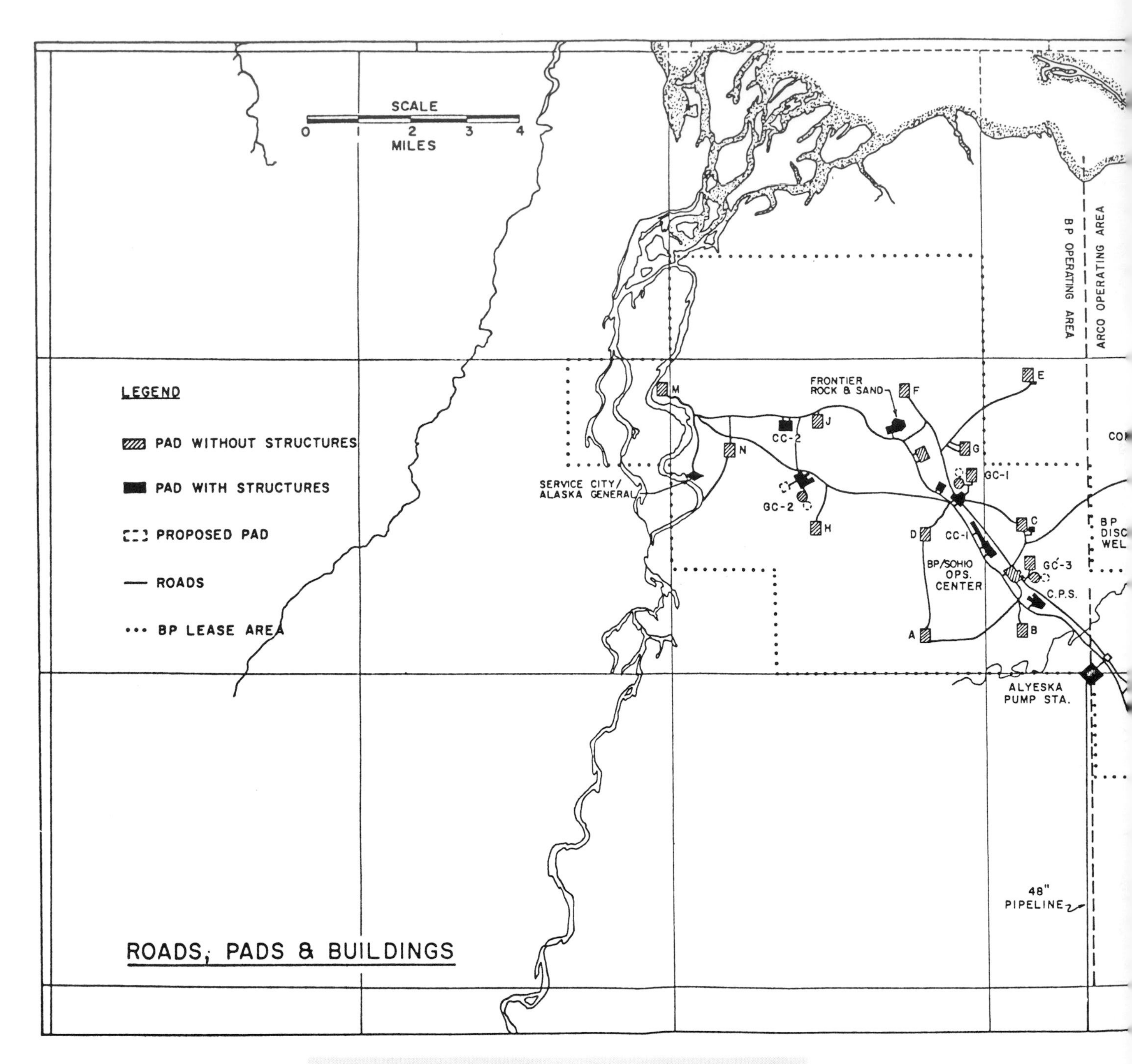

A typical oil drilling rig in the Prudhoe Bay area.
COURTESY ARCO ALASKA

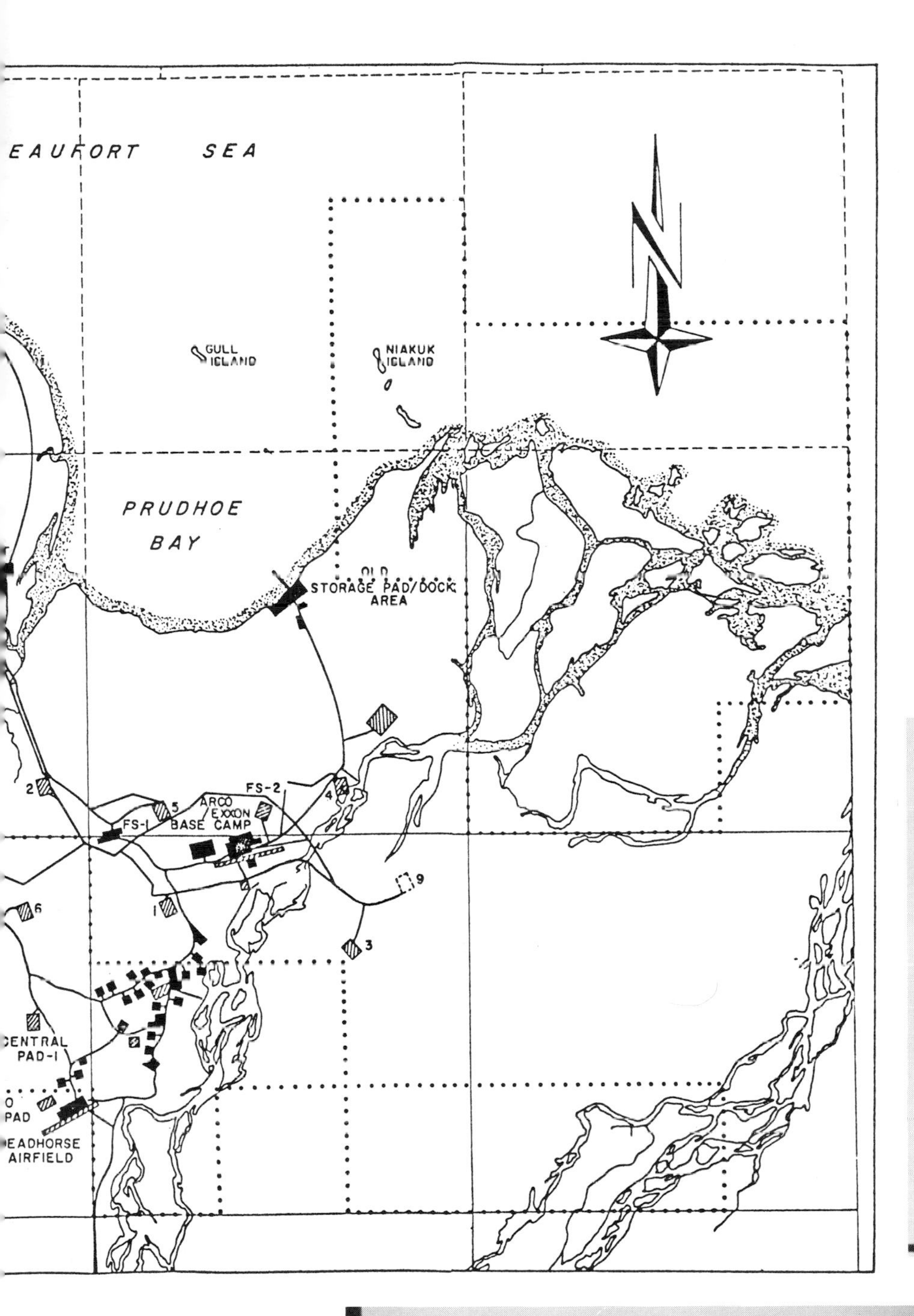

PRUDHOE BAY STATE NO. 1
ATLANTIC RICHFIELD · HUMBLE

THIS WELL, THE DISCOVERY WELL OF THE PRUDHOE BAY FIELD, WAS SPUDDED AS A JOINT VENTURE OF ATLANTIC RICHFIELD COMPANY AND HUMBLE OIL & REFINING COMPANY. THE DISCOVERY WAS ANNOUNCED BY THE TWO FIRMS IN MARCH 1968 WHEN THIS WELL FLOWED ON TEST AT A RATE OF 1152 BARRELS OF OIL AND 1.32 MILLION CUBIC FEET OF GAS DAILY.

THE PRUDHOE BAY STATE NO. 1 WELL, ALONG WITH THE CONFIRMATION WELL ON THE SAGAVANIRKTOK RIVER, SOME SEVEN MILES TO THE SOUTHEAST, DEFINED THE FIRST COMMERCIAL OIL DISCOVERY ON ALASKA'S NORTH SLOPE.

INITIAL QUANTITY OF RECOVERABLE RESERVES IN THE PRUDHOE BAY FIELD WERE ESTIMATED, BY A LEADING CONSULTING FIRM, TO BE 5 TO 10 BILLION BARRELS OF OIL.

A tubular monument to Prudhoe Bay State No. 1.
COURTESY ARCO ALASKA

2. How to get the crude from Prudhoe Bay to the Lower 48

■

HOW TO GET the crude from Alaska's Prudhoe Bay to the Lower 48 was the burning issue in the oil business during the summer of 1968. One proposal was to route giant tankers through the Arctic Ocean around Alaska and south to the West Coast. But this idea was abandoned when it was decided that it would be only a matter of time before a tanker broke up on ice and inflicted a catastrophic oil spill on Arctic waters. (In August 1969 and the spring of 1970 the *S.S. Manhattan*, a 115,000-ton, 1,005-foot-long tanker, made two trips to Prudhoe Bay via the Canadian and Alaskan Arctic to see if a Northwest Passage was feasible. This trial run was jointly sponsored by Humble, ARCO, BP and the Canadian government. This run showed that while it was feasible, it was not economically practicable to use ships.)

Clearly, a pipeline was the only solution.

Pipelines, a common conveyance in the Lower 48, had been built in cold climates before, although never over such a great expanse of tundra, muskeg, mountains, river systems and active earthquake zones. And no other pipeline route passed through such a remote area. The only project that came even close to dealing with the problems the Alaska Pipeline would encounter was the World War II CANOL (Canadian Oil) pipeline, a four-inch pipe stretching some 500 miles from the MacKenzie River in Canada's remote Northwest Territories to Whitehorse in the Yukon.

In Texas, the center of America's oil business, engineers were beginning to grasp the problems of building the world's largest oil pipeline through one of the planet's most inhospitable lands. Experts from many disciplines were hired to pore over every facet of the design and construction. Crews were placed in the field to study the soil and rock conditions that the line would encounter. Even the Russians were consulted, but they had never attempted anything approaching the magnitude of this project.

Before the business of choosing a route began ports were looked at all over the Gulf of Alaska from Cook Inlet to Cordova. Whittier, Seward and Valdez were considered but in the end Valdez won out—it was ice-free and the closest point to Prudhoe Bay. The port at Valdez had been devastated by the 1964 earthquake and had been rebuilt several miles from the original townsite. But a large bedrock bench was located near the townsite that would serve as the terminal site.

There was no getting around the three mountain ranges the pipeline would meet on its way south. The Brooks Range lay in the extreme north, the lower Alaska Range in the middle of the state and the Chugach Mountains along the southern coast. It was obvious that using the low passes through these mountains would reduce construction costs and the amount of pumping stations needed. However, the lower pass in the Brooks Range, Anaktuvuk Pass, was bypassed in favor of Antigun Pass to the east because of more favorable soil conditions and gravel deposits, even though at 4,790 feet, Antigun was double the elevation of Anaktuvuk. From the Brooks Range the proposed route would head south to the next large obstacle, the Yukon River, and then swing east of Fairbanks to cross the Alaska Range over Isabel Pass at 3,500 feet. From there it would follow the existing Richardson Highway through the Copper River Basin to the Tonsina River and then cross 2,500-foot Thompson Pass in the Chugach Mountains to Valdez via Keystone Canyon.

Before any other planning could proceed engineers had to decide whether to build a hot or cold transmission line. In temperate climates, oil is pumped from the ground and transported at its normal temperature (as high as 180 degrees Fahrenheit) with no problems. But Prudhoe oil would be crossing miles of permanently frozen ground called permafrost. Hot oil, made even hotter by friction as it's transported, would melt the permafrost, and the thawed ground would buckle or break the pipe.

A cold-oil pipeline was considered, but found to be impractical because gargantuan reservoirs would be needed to cool the oil before transporting it through the pipe, and vast sums of money would have to be spent to keep the oil cooled to between 30 and 35 degrees fahrenheit for the long trip south.

As soon as it was determined that the Prudhoe Bay

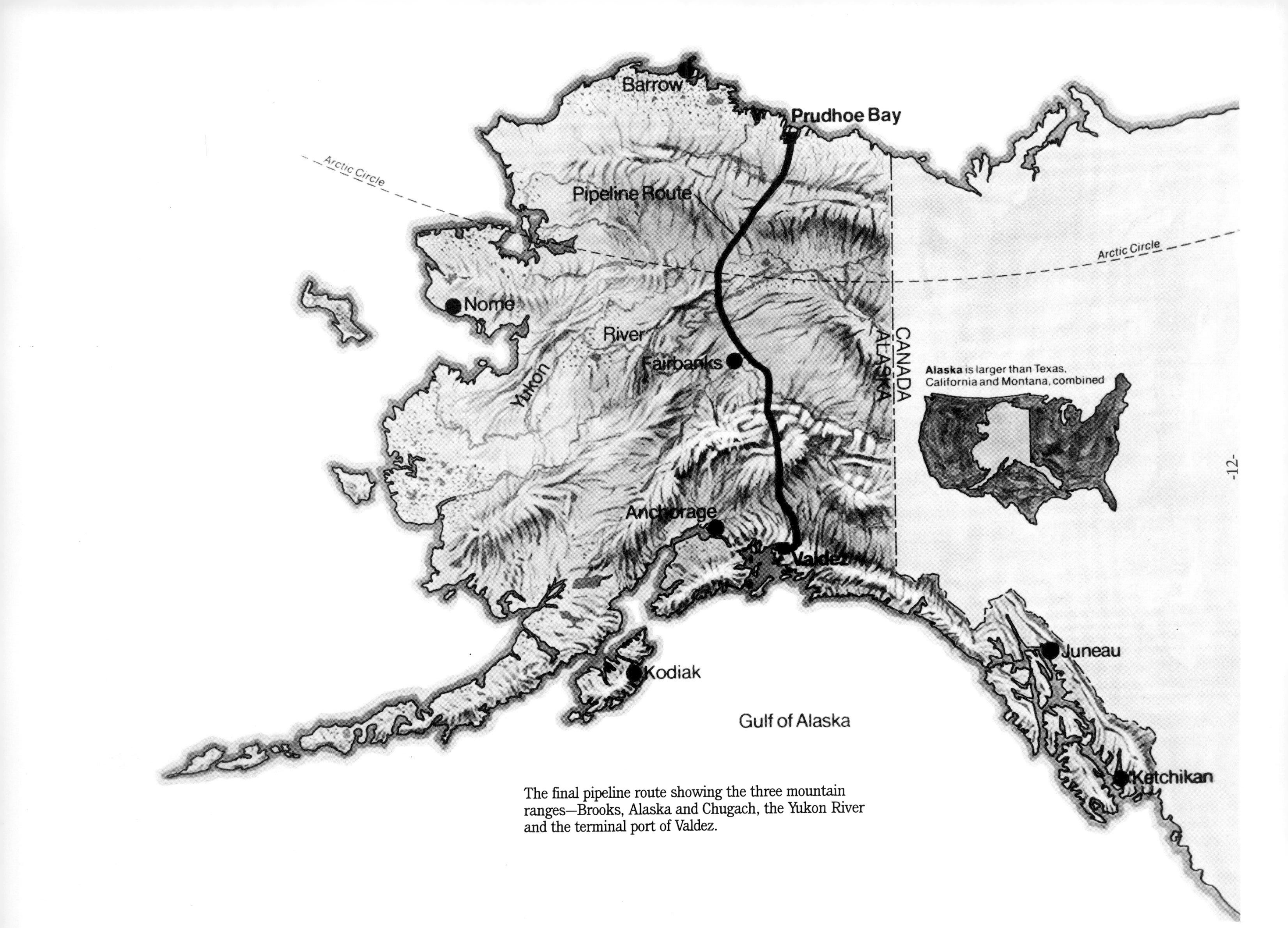

The final pipeline route showing the three mountain ranges—Brooks, Alaska and Chugach, the Yukon River and the terminal port of Valdez.

discovery was a major find, the oil corporations began a frantic scramble for a share of the wealth. Atlantic Richfield Company (ARCO) and Humble Oil & Refining Company (now the Exxon Corporation) came up with the idea for the pipeline and they were joined in 1968 by the British Petroleum Company Ltd. (BP) to form TAPS—The Trans-Atlantic Pipeline System. The next year five more companies joined the consortium: Union Oil Company of California, Phillips Petroleum Company, Amerada Hess Corporation, Mobil Oil Corporation and Home Oil Company of Canada (which later dropped out).

TAPS was succeeded in 1970 by the Alyeska Pipeline Service Company. Alyeska, an ancient Aleut word for Alaska that meant "the Great Land," was established as a non-profit corporation to handle all design, construction and operation of the pipeline. In 1971, Standard Oil Company of Ohio (SOHIO) came into the project by trading 25 percent of its stock for BP's North Slope oil leases. It was now the largest owner of Prudhoe oil.

On February 10, 1969 it was announced that a 48-inch, 800-mile pipeline would be constructed from Prudhoe Bay to Valdez. Construction would begin in the spring of 1970 and be completed in 1972. The cost was projected at $900 million. Its initial capacity would be 500,000 barrels a day and work up by 1980 to two million barrels a day.

All this, of course, lay on the drawing board. All of the geophysical work, which would reveal the need for costly aboveground construction over half the route, and all the political and sociological problems were yet to come. These, and many other factors, would conspire to delay construction. And in the end the Alaska Pipeline would cost eight billion dollars.

The University of Alaska at Fairbanks, was one of the prime sites for research carried out for the pipeline. Six hundred feet of warm pipe was buried in the permafrost, as shown here, to verify Alyeska's mathematical models describing heat flow and permafrost melting and to test the pipeline's effect on vegetation.

Exposed permafrost along the pipeline is examined by a geologist. Permafrost is defined as "unconsolidated deposits of bedrock that continuously have had a temperature below zero degrees centigrade for two years or more." In areas where the permafrost is "ice-rich" or otherwise thaw-unstable, the pipeline is elevated to avoid thawing the ground and thus losing support for the pipe.

Slope Camp at Prudhoe Bay was leased by Alyeska Pipeline Service Company to house pipeline construction workers. About 65 percent of the 165 miles of pipe stored in the yard was transported to locations along the pipeline route by the end of July 1975.

A specimen of the 48-inch pipe to be used in the trans Alaska pipeline is shown being hoisted into position in the four million pound testing machine at the structural laboratory of the Engineering Department at the University of California, Berkeley, where pipe strength tests were conducted for Alyeska Pipeline Service Company.

3. On again, off again. Will the pipeline be built?

IN 1968 THE ROAD north from Fairbanks ended 80 miles away at Livengood—smack dab in the middle of the Alaskan wilderness. From here, it was 360 miles of forest, mountain and tundra to Prudhoe Bay—and the oil. Before a single length of pipe could be laid a road would have to be built along this wild right-of-way. (Walter J. Hickel, Alaska's progressive governor, had proposed a temporary route some years before to haul supplies and equipment to the North Slope. Under his leadership the state of Alaska built a winter road from Livengood to Prudhoe. The "Hickel Highway" was carved out of ice and was only usable in the winter. When the spring thaws came the corridor became a ribbon of mud.)

Still, an all-weather road would have to be constructed. Although that project was scheduled to begin in 1969 it was delayed—like everything else associated with the pipeline. In June 1969 TAPS applied for a right-of-way for the road and the pipeline. Finally, in August, Hickel, now the U.S. Secretary of the Interior, conferred with Congress and permitted TAPS to build a section of the road from Livengood to the south bank of the Yukon River. It was a well-engineered project and enabled TAPS people to gain valuable experience for the gargantuan task of building the pipeline itself.

In December, an ice bridge was built over the Yukon River in anticipation of moving heavy equipment north for construction the coming spring. Tundra Contractors Inc., a Fairbanks outfit owned by Alaskan natives, built this bridge by laying logs over the ice and pumping water over them to form a road bed on an ice base.

There was a frenzy of activity in the winter of 1970 as TAPS prepared to accept federal permits for access to the government lands over which the pipeline would cross. By spring, supplies, equipment and manpower all had to be in place along the route. This could only be accomplished in the winter when travel over the frozen ground was possible. TAPS had taken a risk in 1969-70 and dropped supplies and some construction equipment along the route and even constructed camps for the workers, in anticipation of being allowed to begin.

But the first piece of pipe wouldn't be laid until 1975.

Part of the planning going forth in 1968 for the 1969 season was the ordering of 800 miles of 48-inch pipe. This had to be done even before all the permits were gathered in order to get it dropped along the route so construction could commence in 1970.

Finding this much pipe would not be easy. Specs were put out to steel mills in the United States, Britain, Germany and Japan. The pipe TAPS needed was peculiar: Its metallurgical characteristics had to be such that it could withstand the conditions. It had to be compliant enough to yield to the enormous changes in temperate characteristic of this part of the world. In other words, it was a very specialized order. And TAPS wanted it right away.

The only mills that would agree to these requirements were Japanese. In 1969 three companies started production on a $100 million order. Delivery of the first product was scheduled for September 1969. Thousands of 60-foot sections of pipe would be barged from Japan through the Bering Sea and around the top of Alaska to Prudhoe Bay, or as 40-foot sections shipped to the southern terminus of Valdez and to Anchorage to be put on the Alaska Railroad for the trip north to Fairbanks. At Valdez and Fairbanks the pipe sections would be welded together and distributed by trucks up and down the pipeline route. This would save time and further reduce the exposure of workers to the elements.

The Big Chill

In 1969 Congress passed the National Environmental Policy Act, which required that an environmental impact statement (EIS) and a study of alternatives to projects such as the pipeline be submitted before any construction on the pipeline could proceed. In that same year Secretary Hickel allowed a freeze to continue that had been placed in 1966 on land transfers until native American territorial claims against the

Miles and miles of 16-foot long, 48-inch pipe is stockpiled at Prudhoe Bay waiting to be hauled south for installation.

North Star Storage Area in Fairbanks.

By the end of July 1975, approximately 145 miles of pipe had been transported from the Fairbanks pipe storage yard. About 240 miles of pipe had been stored in the yard since 1970.

federal government could be settled.

Permits granted to TAPS for passage over several native lands were withdrawn and lawsuits were initiated against the Secretary of Interior, and that halted any further movement. Several environmental groups also filed lawsuits to stop further work until many environmental questions could be answered.

And to further complicate matters there were still many important technical questions to be answered. The most pressing of these was whether to bury the pipe or not. Scientists and engineers still didn't know everything they would have to about the soils along the intended route of the pipeline. Wildlife was also an important consideration. What would an above-ground pipeline do to the migration patterns of caribou? What would be the impact on wildlife from the construction process and later from the operation of the pipeline?

It would take several years, hundreds of hours of testimony, and thousands of pages of reports to convince the government to grant all the permits needed. The $900 million construction figure that was projected in 1968 was a dim memory. With inflation, the decision to construct some of the pipe aboveground, and all the delays, the price tag had jumped in 1970 to $3.5 billion. The 800 miles of pipe that had cost $100 million in 1969 now looked like a real bargain.

Ft. Wainwright, near Fairbanks, was used as the pipeline headquarters for the northern construction projects.

Pipe is stacked in the storage yard.

The Final Go Ahead

In 1971 the first draft of the EIS was issued and hearings on the pipeline were held in Anchorage. In 1971 Richard Nixon fired Walter Hickel. Hickel was regarded as pro-pipeline although he gave the environmentalists a chance to speak while be was in office. And they had a lot to say. During this period countless other alternatives for moving the oil south were proposed, studied and rejected.

These schemes ranged from a fleet of oil-laden submarines, to extending the Alaska Railroad north to building a super highway and trucking the oil south. None of these schemes was practical. A major pipeline was still the only real alternative.

Even while the hearings over native land claims continued during The Big Chill, engineering and field studies were pushed forward to find the least expensive, safest and shortest pipeline route south. And soils tests were still being made to determine whether to elevate or bury pipe in problem soils called thaw-unstable permafrost. Then there was Alyeska's plan to bury the pipeline under the wide and often treacherous Yukon River. This was not an overwhelming

President Richard M. Nixon signing the Alaska Pipeline Authorization Act on January 23, 1974. From left: Sen. J. Bennett Johnson (D-La.), Sen. Mike Gravel (D-Alaska), Sen. Henry M. Jackson (D-Wash.), Rogers C.B. Morton, Secretary of the Interior, Sen. Ted Stevens (R-Alaska), Rep. James A. Haley (D-Fla.), Rep. John Melcher (D-Mont.), Rep. Harold T. Johnson (D-Calif.), Sen. Paul J. Fannin (R-Ariz.), Rep. Don E. Young (R-Alaska), Rep. Craig Hosmer (R-Calif.), Rep. Morris K. Udall (D-Ariz.), Sen. Clifford P. Hansen (R-Wyo.) and Sen. Mark Hatfield (R-Ore.).

engineering challenge, but the state of Alaska decided to build a permanent bridge across the river and hoped that by strapping the pipe on the side of the bridge this would help pay construction costs.

The state also looked into taking over construction of the pipeline and thus owning it. There were no legal precedents for this move and with the inflated price for engineering and construction now pushing past the $3.5 billion mark, the state backed out of the discussion.

More lawsuits were filed against the pipeline during the early 1970s, causing further delays. Alaska itself was becoming a victim of these delays; it was losing millions of dollars in royalty payments on all the oil which was still in the ground waiting to be conveyed to Valdez.

One of the biggest legal hurdles was settled, however, late in 1971. The Alaska Native Claims Settlement Act gave the natives the right to select 44 million acres of land and receive $462 million over an eleven-year period plus a two percent mineral royalty until an additional $500 million was paid. For all this the natives gave up any further claims to land in the state. Twelve regional native corporations were established to oversee this large settlement of land and money. This act lifted the land freeze, enabling the pipeline to move a little closer to actual construction.

Additional proposals came along in 1972 for alternative routes. The most popular of these was a 2,600-mile route across northern Alaska to Canada's MacKenzie River area and then south to Alberta to join existing pipelines. There were many arguments for and against this route but huge costs, national interests and possible severe environmental problems in the Arctic National Wildlife Refuge finally put this proposal to rest.

In 1972 Secretary of the Interior Rogers C.B. Morton approved right-of-way permits for the pipeline. But environmental groups continued to press their lawsuits. One of their arguments was the fact that Alyeska had requested a right-of-way wider than the fifty-foot limit imposed by the 1920 Mineral Leasing Act.

Work continued on the pipeline's design through 1972. A concensus was forming that more than half of the pipe should be elevated over permafrost areas, instead of buried in the more conventional way.

In 1973 the pipeline question, after being hustled in and out of federal courts for years, finally wound up on the floor of the U.S. Congress. And international events late that year forced the issue. In October, Egyptian military units attacked Israel. Israel retaliated, and the result was the Yom Kippur War, a total embargo of Middle Eastern oil to the United States, and lines at gas stations stretching around the block. Clearly, for reasons of national security, the U.S. was going to need its own supplies of oil.

Congress granted permission for right-of-way. An amendment to the bill sponsored by Alaska's Senator, Mike Gravel, put a time limit on further legal delays in the acceptance of the EIS. The pipeline finally got the go-a-head. Gravel's amendment led to a 49-49 tie in the Senate which was broken by Vice-President Spiro Agnew. The pipeline had finally been cleared to proceed. The Trans Alaska Pipeline Authorization Act of 1973 was passed by both houses of Congress and signed into law by President Nixon on November 16, 1973.

At last, construction could begin on one of the largest human endeavors ever attempted. It would alter the U.S. oil industry forever and fatten the coffers of the Alaska State Treasury for years to come.

Anchorage Daily Times

SENATE OK'S PIPELINE AS AGNEW BREAKS TIE

Jackson Bill Goes With Added Punch

Senate Nod Came Today In 5 Steps

House Panel Okays Bill

Inside

Alaskans Laud Nod For Oil Pipeline

Weather

Borough To Vote On 11-District Rep

Anchorage Daily Times

After Five Years, Two Months And 19 Days, It's Official:

PIPELINE WORK BEGINS

Alyeska Assigns 1200 Men To Road

Unions Schedule Fairbanks Meet

Tape Plan Forwarded

'We Spent Too Much Money'

Weather

Stans, Mitchell Still Face Hurdles

Session Focuses On Child Rights

On The Inside

Building the Pipeline Haul Road

First of new camp buildings arrives at storage yard at Fairbanks after transport from Calgary, Alberta. The structures were moved to remote construction camp sites north of the Yukon River, to house crews constructing the highway from the Yukon River to Prudhoe Bay. The camp buildings, purchased by Alyeska from Atco Structures, Inc., of Anchorage, a subsidiary of Atco Structures, Ltd., of Calgary, were part of a total order of 422 buildings delivered in Fairbanks by Atco. The camp structures were pulled by truck tractors from Calgary. Yard equipment of H&S Warehouse, Inc., and Earthmovers of Fairbanks, joint venture, was used to move units within the storage yard.

One of the larger bridges along the haul road north of the Yukon River is this one across the Dietrich River.

Climbing the Chandelar Bench in the Brooks Range.

An equipment train on the old Hickel Highway.

Alyeska pipeline
"First OFFICIAL VEHICLE TO DRIVE THE ALL WEATHER HIGHWAY
FROM THE YUKON RIVER TO PRUDHOE BAY ALASKA"

Alyeska pipeline
SERVICE COMPANY
ROAD LINK-UP SITE, SOUTH FORK KOYUKUK RIVER VALLEY
YOU ARE 100 MILES NORTH OF THE YUKON RIVER AND 32 MILES INSIDE THE ARCTIC CIRCLE
ON THIS SITE, SEPTEMBER 29, 1974, THE FINAL LINK WAS COMPLETED IN THE FIRST OVERLAND ALL-WEATHER ROUTE FROM THE YUKON RIVER, IN ALASKA'S INTERIOR TO THE ARCTIC COAST, 360 MILES NORTH, AT PRUDHOE BAY.
THE DISTANT POINTS WERE JOINED JUST 154 DAYS FROM THE SIGNING OF CONSTRUCTION CONTRACTS ON APRIL 29.1974.
ABOUT 32 MILLION CUBIC YARDS OF GRAVEL WERE REQUIRED TO COMPLETE THE ROAD, WHICH ALSO INCLUDES 20 BRIDGES OVER STREAMS, AND A MOUNTAIN CROSSING AT A 4,500-FOOT-HIGH PASS IN THE BROOKS RANGE. AT THE PEAK OF CONSTRUCTION, SOME 3,300 WORKERS WERE EMPLOYED IN THE ROAD WORK.
THE ROAD CONSTRUCTION WAS THE FIRST WORK UNDERTAKEN IN THE TRANS ALASKA PIPELINE PROJECT. THE PIPELINE WILL TRANSPORT THE 9.6 BILLION BARRELS OF PROVEN CRUDE OIL RESERVES AT PRUDHOE BAY, 798 MILES SOUTH TO THE ICE-FREE PORT OF VALDEZ. THE OIL WILL THEN BE SHIPPED IN MARINE TANKERS TO PORTS ON THE WEST COAST OF THE UNITED STATES.
CONSTRUCTION MANAGEMENT CONTRACTOR:
BECHTEL INCORPORATED
PROJECT DESIGN ENGINEERS:
MICHAEL BAKER, JR.
EXECUTION CONTRACTORS:
GREEN/ASSOCIATED PIPELINE
MORRISON-KNUDSEN COMPANY, INC.
GENERAL-ALASKA-STEWART
BURGESS CONSTRUCTION CO.
BARROW
Prudhoe Bay
Proposed Pipeline Route
NOME
FAIRBANKS
ANCHORAGE
VALDEZ
JUNEAU
KODIAK
KETCHIKAN
THE 798 MILE ROUTE OF THE PROPOSED TRANS ALASKA CRUDE OIL PIPELINE EXTENDS FROM THE NORTH SLOPE OIL FIELDS TO THE ICE-FREE PORT OF VALDEZ.

FINAL ROAD LINKUP
SOUTH FORK KOYUKUK RIVER
COLDFOOT - PROSPECT
PRUDHOE BAY - YUKON RIVER
SEPTEMBER 29, 1974

Construction Camps

During the course of construction, 29 camps, 19 of them major sites, were built along the 800-mile route of the pipeline. Fifteen of these were located north of the Yukon River, where work conditions were harsher than in the southern sections. These camps, while temporary in nature, nevertheless boasted the ultimate in modern facilities to accommodate the thousands of men and women working in some of the most adverse weather conditions in the world.

In order to keep the construction process on schedule the camps had to be able to solve any problem in supply, equipment repair or personnel that might crop up. Although some of the camps were located along or just off the all-weather Richardson Highway and had access to the outside world, the more remote camps —especially those north of the Yukon River—had to be self-sufficient.

The camps were a very important and integral part of the entire construction process.

The Franklin Bluffs camp, 40 miles south of Prudhoe Bay.

The Toolik camp was 130 miles south of Prudhoe Bay and 10 miles north of the Galbraith Lake camp.

A C-130 Hercules transport plane unloads its supply of fuel into rubber "pillow" tanks at Happy Valley.

Happy Valley camp, 90 miles south of Prudhoe Bay, was actually two camps in one area. Its airfield is pictured.

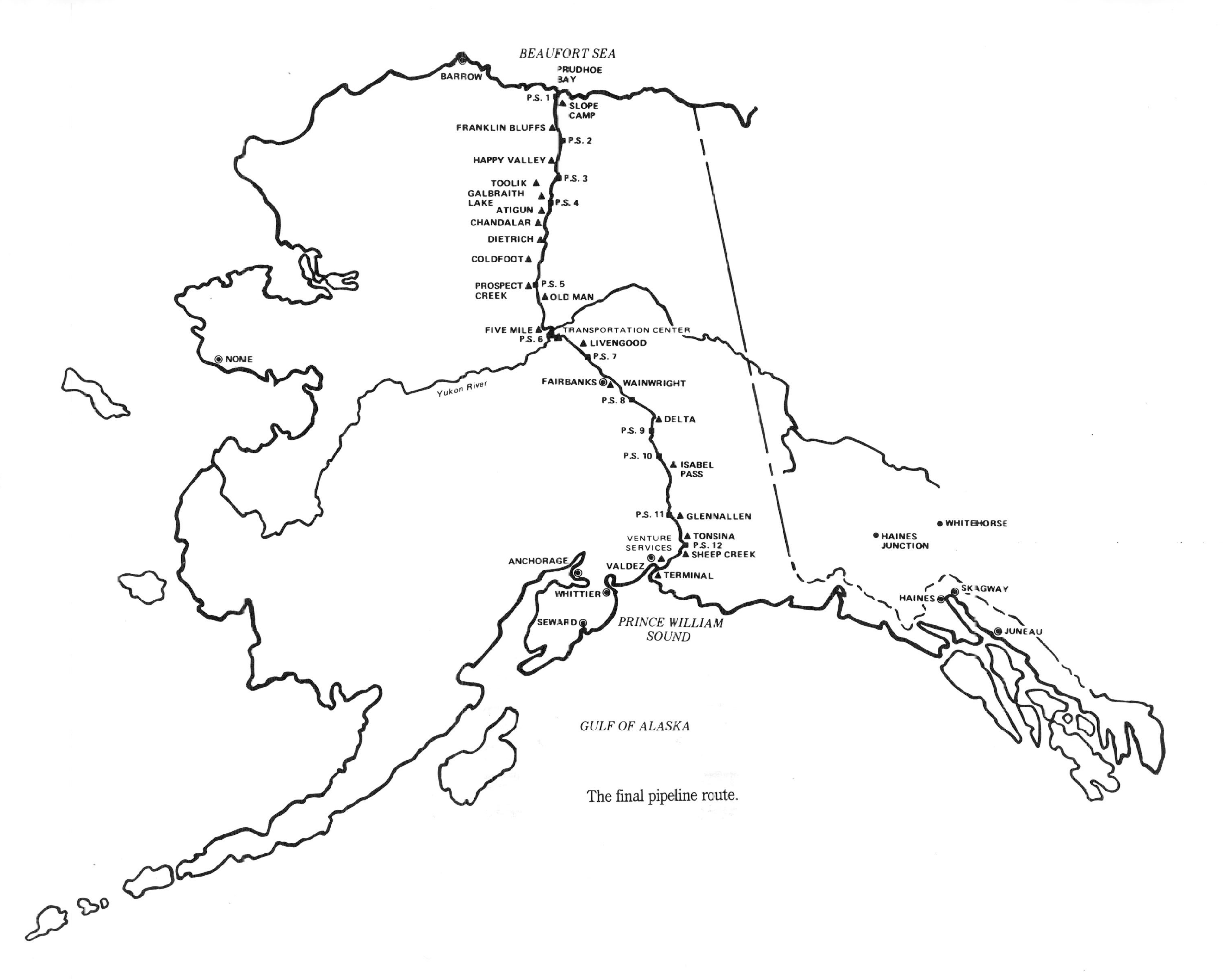

The final pipeline route.

The Galbraith Lake camp was located at the foot of the Brooks Range 140 miles south of Prudhoe Bay.

The airfield at Galbraith Lake was a mile-long landing strip, equipped with lights and navigational aids. The FAA approved the airport for instrument approaches.

A look at the extreme isolation of some of the northern construction sites: This is an early view of Galbraith Lake.

The Chandalar camp was located three miles south of the continental divide in the Brooks Range. In July 1975, camp occupancy was about 75.

The Atigun camp was located north of the continental divide in the Brooks Range, three miles from Atigun Pass. It was perched 4,800 feet about sea level and was the highest of the pipeline's camps.

The Dietrich camp was located on the Dietrich River in the southern foothills of the Brooks Range, 200 miles south of Prudhoe Bay. In July 1975, camp occupancy was about 550.

The Coldfoot camp was located 240 miles south of Prudhoe Bay.

The Prospect Creek camp was located 80 miles north of the Yukon River and 16 miles north of the Arctic Circle. The Jim River ran alongside the camp. Prospect Creek flowed nearby. It could accommodate 600 workers.

The Old Man camp was located 52 miles north of the Yukon River. It was headquarters for Pipeline Section 4 and had an occupancy of 875 in July 1975.

The Five Mile camp was located five miles north of the Yukon River and 350 miles south of Prudhoe Bay. In July 1975 the occupancy was about 570.

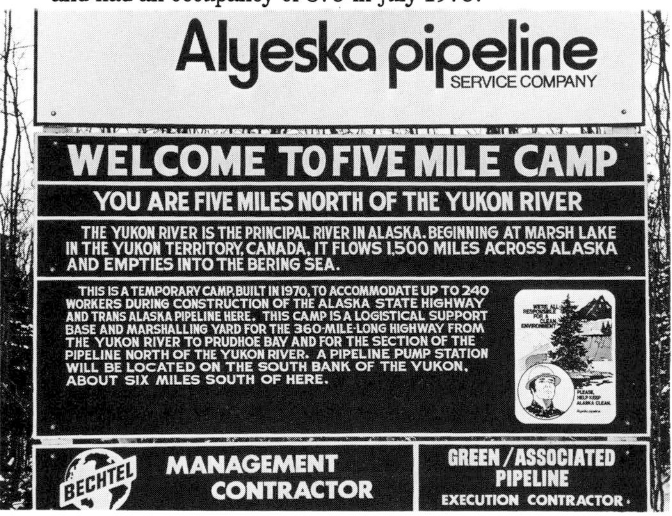

The first camp south of the Yukon River was located at Livengood, 60 miles north of Fairbanks.

The Delta camp was located at the Tanana River crossing, 80 miles south of Fairbanks.

The Isabel Pass camp was located north of the pass it was named for in the Alaska Range, about 600 miles south of Prudhoe Bay.

The Tonsina camp was located 68 miles north of Valdez. The camp took care of 1,232 workers during the peak of construction. The triangular-shaped structure in the foreground is a holding pond for the camp sewage treatment plant. At top left is a 3,000 foot airstrip. The road running diagonally through the photo is the Richardson Highway.

The Glennallen camp was located near Glennallen, a town about 110 miles north of Valdez. It was the headquarters camp for Pipeline Section 1 and had an occupancy of about 875 in July 1975.

The Sheep Creek Camp was two miles north of Keystone Canyon off of the Richardson Highway, 20 miles north of Valdez. Thompson Pass through the Chugach Mountains is at the top. The Sheep Creek Camp's occupancy in July 1975 was 350.

Bladder storage tanks were set up at the camps to provide the thousands of gallons of fuel needed to run the camps and the construction equipment.

Camp living conditions were exceptional. For morale reasons, they had to be. Workers put up with remoteness, long working hours and extreme weather conditions.

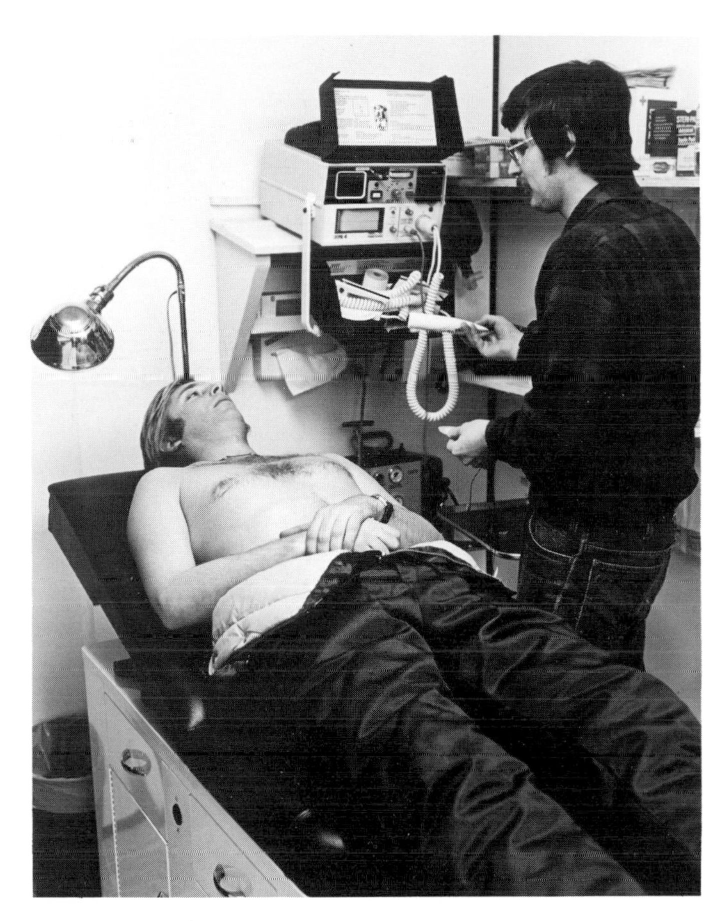

A Field Medical Technician (FMT) at Galbraith Lake camp prepares to administer an EKG to a worker. The electrocardiograph could be read over the telephone to a physician in Fairbanks. FMTs were certified to set fractures, perform minor surgery and administer most drugs. Medical facilities at the construction camps included a waiting room, examination room and four bed ward.

Recreational facilities, both indoors and outdoors were first class. It was the policy of the Alyeska Pipeline Service Company to provide the best accommodations, recreation and cuisine that was possible under the conditions of the Alaskan wilderness.

PIPELINE CONSTRUCTION SECTIONS & CONTRACTORS

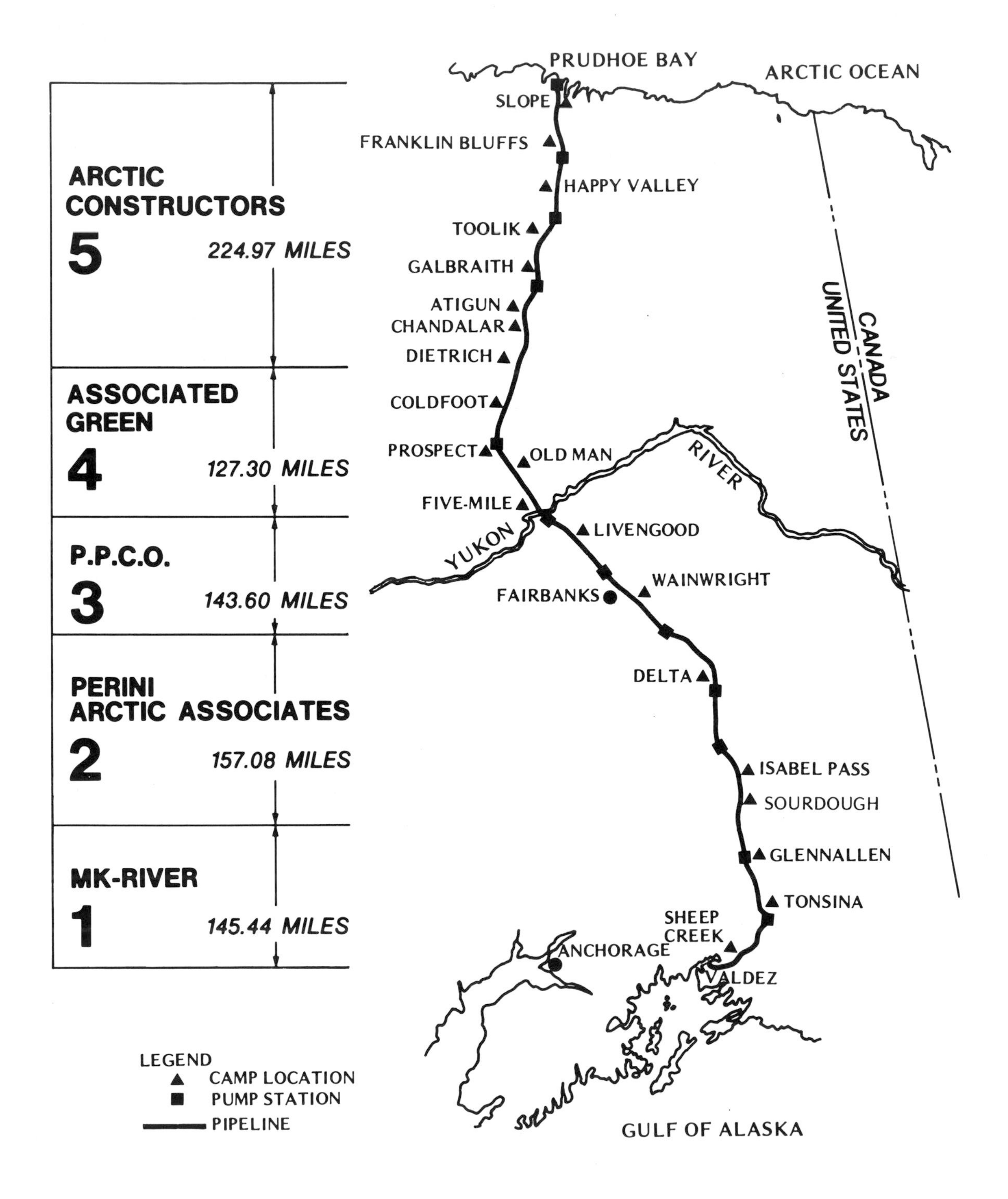

4. "We didn't know it couldn't be done." The construction process

Text courtesy of Alyeska Pipeline Service Company.

OPERATION ACTIVITIES on the trans-Alaska pipeline center largely in pump stations and the Terminal. The 800 miles of pipe which carry the oil from Prudhoe Bay to Valdez play a purely passive role. However, design and construction of the pipeline involved a number of complex problems and unique solutions.

Most pipelines are buried below ground. But because of permafrost conditions across Alaska, more than half of the trans Alaska line is built above ground and several short buried sections are specially refrigerated.

The method chosen for emplacement of pipe along the route depended, for the most part, on soil conditions and the effects of pipeline heat on the soil. In stable soils such as rock, in thaw-stable sands and gravel, or in conditions where thawing would not cause unacceptable disruption of terrain, the line is buried in a conventional manner. The pipe was placed on a layer of bedding material in a ditch eight to 35 feet deep and then covered with prepared gravel padding and soil fill material.

In places where melting permafrost might create difficult soil stability conditions, the pipeline was built above ground. In this mode, the pipe is supported by crossbeams installed between vertical supports placed in the ground. The pipe itself, insulated with 3¾ inches of fiberglass and jacketed with galvanized steel, is mounted on Teflon-coated shoe assemblies which can slide on the crossbeams.

Thawing around the vertical supports is prevented by thermal devices, called heat pipes, which are installed as necessary inside the supports. Inside the self-contained heat pipes is a refrigerant which vaporizes below ground, then rises and condenses in the above-ground radiators, removing ground heat whenever the ground temperature exceeds the temperature of the air.

To allow for contraction and expansion of the above-ground pipe because of temperature changes (temperatures before start-up ranged down to minus 70 degrees Fahrenheit, but oil temperatures may reach as high as 145 degrees), above-ground sections were built in a flexible trapezoidal zigzag configuration. In this design, longitudinal expansion of the pipe is converted into a sideways movement. The configuration also accommodates pipe motion induced by an earthquake.

Anchor structures erected every 800 to 1,800 feet hold the pipe in position. Between anchors, however, the pipe can move sideways on the crossbeams as much as 146 inches due to thermal expansion and contraction and an additional 24 inches due to seismic activity. At points where potential earthquake movement might be larger, bumpers have been installed on supports to limit horizontal movement of the pipe and absorb the energy of its impact.

At the Denali Fault, the only geologically active fault crossed by the line, design permits as much as 20 feet of horizontal and five feet of vertical motion.

At one highway crossing and two points where caribou migration routes cross the pipeline in permafrost soils, the pipe is buried in a refrigerated ditch. Refrigeration plants at each point circulate chilled brine through loops of six-inch pipe to maintain soils in a stable, frozen condition.

At more than 800 river and stream crossings, the pipe either bridges the waterway or is buried beneath it. At most small streams, the elevated pipe bridges the water on conventional supports. At 14 places, however, special bridges were built. Pipe across the Yukon River is attached to a highway bridge. Standard plate girder structures were built at 10 crossings. Special suspension bridges were constructed across the Tanana and Tazlina rivers, and a tied-arch bridge was designed for the Gulkana River.

A total of 151 gate and check valves were placed in the line. Eighty-seven valves protect streams, 10 protect population areas and three protect sites where the environment is especially sensitive. Spacing of the valves limits the size of any potential leak.

Check valves, designed to be held open by flowing oil and to close automatically when oil flow stops or is reversed, prevent the reverse flow of oil on uphill sections of the line should oil leaks occur. To increase pipeline operating efficiency, some check valves are held fully open mechanically, thus lifting valve components entirely free of the oil stream. These valves are fitted with "actuators," which sense pressure changes in the event of reversed or stopped oil flow, and cause the valves to close. Gate valves, generally on flat terrain and downhill slopes, isolate sections of the line, and, thus, limit spills. Sixty-one of the 71 gate valves can be closed and opened by the Pipeline Controllers in Valdez. All valves can be operated manually for maintenance of the line.

The pipe itself was specially engineered and manufactured for the trans Alaska line. It was manufactured in three grades, with minimum yield strengths of 60,000, 65,000 and 70,000 pounds per square inch, and in two wall thicknesses: .462 and .562 inch.

The pipe was coated and wrapped for protection from bacteriological, chemical and electrolytic corrosion. In buried modes, zinc ribbons buried with the pipe serve as sacrificial anodes. They prevent a flow of current away from the pipe which would corrode its metal surface.

However, even this overlapping corrosion prevention system cannot be 100 per cent effective for buried pipe over many years of pipeline operation. Thus, as a part of a comprehensive monitoring program, the pipeline is checked periodically for corrosion, and corrective measures are taken if necessary.

The monitoring program extends, also, to pipe settlement. In addition to the electronic pig used to detect pipe deformation, special monitoring rods are employed to directly check for any settlement or other movement of the buried pipeline. These metal rods are welded to the top of the pipe at periodic locations, and extend to several feet above the ground surface. By surveying the elevations of the rods at regular intervals, monitoring crews can readily detect any signs of foundation settlement or shifting.

Where unusual stresses are noted, the buried pipe may be excavated and examined. Repairs are made when necessary.

This construction culminated with the first flow of oil to the south on June 20, 1977.

■ The Yukon River Bridge

One of the major obstacles to overcome on the haul road construction project was crossing the mighty Yukon River. In its entire length of 2,000 miles, from its headwaters in Canada to its mouth in the Bering Sea, only one bridge cross it and it is in the Yukon, Canada. It was an obstacle that was finally overcome in October 1975 when the 2,300 foot bridge was opened to truck traffic. With the final link-up of the road at the South Fork of the Koyukuk River Valley traffic could now move from Fairbanks north to Prudhoe Bay.

Ice forms around a cofferdam as winter sets in. The work continued within the dam however, all winter.

After the initial go ahead for the haul road was received in 1974, work started on the Yukon River bridge. Four cofferdams were constructed in the river from which the piers for the bridge were built. Barges were moored on the downstream side of a dam to provide a work platform.

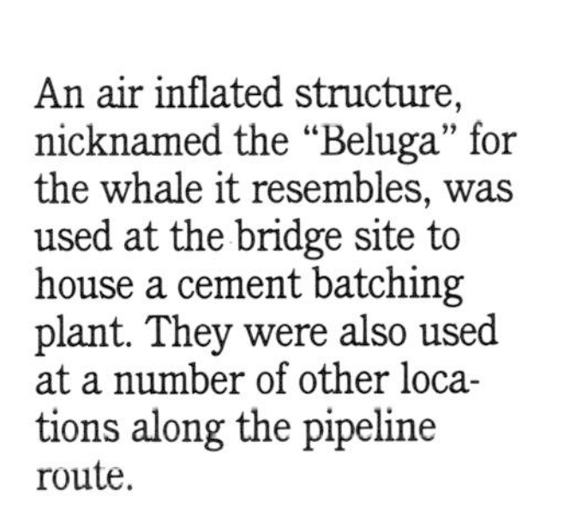

An air inflated structure, nicknamed the "Beluga" for the whale it resembles, was used at the bridge site to house a cement batching plant. They were also used at a number of other locations along the pipeline route.

Progress in the summer of 1975.

Steel girders rest on temporary steel supports on the frozen river on which the permanent piers are constructed.

Progress was made all through the summer of 1975 building the bridge from both sides of the river. It had to be built strong enough to withstand the ravages of the river especially in the spring breakup.

Once the bridge was completed the 48-inch pipeline could be installed alongside the bridge, thus it served two purposes, carrying both vehicular traffic and the oil pipeline.

In early October 1975 the 2,300 foot long Yukon Bridge was opened to traffic. It was built by the State of Alaska and partially financed by the Alyeska Pipeline Service Company.

Ribbon cutting on the new Yukon River Bridge, October 1975.

Archaeological Findings

A Mammoth find was recovered during construction work for the trans Alaska pipeline. The relic, the tip of a mammoth tusk, was recovered by William V. Peterson, shown here holding the ivory piece, at a road construction site about five miles north of the Yukon River. Peterson was employed by Green/Associated Pipeline, a contractor engaged to build a portion of the haul road from the Yukon River to Prudhoe Bay. The relic was turned over to the University of Alaska, for preservation in accordance with the Federal Antiquities Law. The site of the find was inspected by archaeologists, but no additional relics were recovered. Mammoths roamed the earth in late Pleistocene times, from 8,000 to 30,000 years ago. Relics of the large, elephant-like beasts are moderately common in Alaska.

Artifacts, some dating back 12,000 years, were found along the pipeline route. As many as 76 archaeologists at one time were working to preserve remnants of earlier civilizations before pipeline construction crews began work. They labored in mosquito-swarm days of summer as well as the dark, sub-zero days of winter.

Archaeological investigation at Healy Lake. The site is at least 12,000 years old.

Archaeologists from the University of Alaska and Alaska Methodist University sift through the remains of ancient civilizations along the route of the pipeline. Over a period of five years, more than 300 sites were excavated yielding more than a half million stone flakes and artifacts.

While heavy equipment is being used to bury a section of the pipeline, at left, archaeologists from the University of Alaska used garden trowels to uncover artifacts about 50 miles north of the Yukon River. Projectile points and stone tools, believed used by Athabascan people who roamed Alaska's interior 3,000 to 5,000 years ago, were unearthed.

Transportation Facilities

Seattle was a major port of embarkation for the thousands of tons of equipment and supplies headed north by barge.

An inflatable hovercraft was employed to ferry heavy trucks across many rivers on the pipeline route.

Many of Alaska's scheduled and non-scheduled airlines were contracted during the construction period to ferry men and equipment.

Alaska International Air's C-130 at Tonsina.

Large airplanes, including the giant military C-130 transport were contracted to haul equipment to the far northern sites along the pipeline route.

Barge loaded at Seattle for the long trip around Alaska to Prudhoe Bay.

A giant Sikorsky S-64 Skycrane helicopter was used to airlift heavy equipment to the top of Keystone Canyon. The 88-foot long helicopter carried bulldozers, drills and other equipment to otherwise inaccessible places during construction of the pipeline through the canyon.

A fleet of Japanese ships delivered the 800 miles of pipe to ports in Alaska.

A barge loaded with sections of pipe is moored in Port Valdez waiting for a tug to tow it to Whittier. The eight rail cars on the barge will be off-loaded at Whittier and hauled by the Alaska Railroad to Fairbanks. Each rail car is carrying six 40-foot-long sections of pipe.

Pipe is being off loaded at Fairbanks from Alaska Railroad flat cars.

Hundreds of trucks carried the 80-foot "double-joint" sections of pipe along the 800-mile route of the pipeline.

A fleet of trucks ready to deliver their load of pipe.

Trucks were also used to transport the thousands of different items needed during the construction period. This truck is loaded with bales of styrofoam insulation used on the workpads of the pipeline.

From storage sites at Valdez, Fairbanks and Prudhoe Bay trucks delivered pipe to sites along the pipeline route.

Rock drills bore holes for blasting in rock along the pipeline route.

A pair of Caterpillar front-end loaders fill a Euclid B-70 bottom dump truck (also called a belly-dump). The giant truck, with a capacity to haul 70 cubic yards of material, was one of the many types of trucks operating along the pipeline route. The material was used to construct access roads and a 50-foot-wide workpad for pipeline construction equipment.

Boards of polystyrene insulation form a mosaic 50 miles south of Prudhoe Bay. Two to three feet of gravel werew placed on top of the insulation to form the work pad for the equipment that would be used to install the pipeline. The insulation protected the permafrost underneath and reduced the amount of gravel that was required to finish the project.

Sidebooms lower concrete coated pipe into the main channel of the Chatanika River, about 15 miles north of Fairbanks. About 350 feet of the pipe was buried to a depth of 12 feet below the low point of the river's main channel. The length of the entire crossing, including the river's floodplain, totals about about 1,500 feet.

A section of pipe is ready for burial near the Little Tonsina River, 65 miles north of Valdez. Concrete saddles weighing more than 9 tons each are used to sink the pipe to the bottom of the water-filled ditch.

Crushed material is deposited over a section of pipe 50 miles north of Valdez. The backfilling machine picks up the material with an auger and transports it over the ditch on a conveyor belt. The pipe in the water-filled ditch, has been weighted down with concrete saddles.

A bent joint of pipe is moved into position alongside above-ground supports, where the pipe joints will be welded into a long section and lifted atop the supports. The pipe is bent to provide a zigzag alignment, required to permit pipeline movement caused by thermal and other forces.

A hydraulic bending machine is used to bend a section of pipe.

Sideboom tractors lower a pre-bent section of 48-inch-diameter pipe into a ditch adjacent to the Little Tonsina River, north of Valdez. The pipe is bent to conform to the contour of the ditch. Tractor operators and laborers work in unison to assure proper pipe alignment.

These above-ground supports are ready to receive the 48-inch diameter pipe, which was welded beside the supports, then lifted into place.

This section of the pipeline, which crosses an access road 18 miles north of Valdez, is 18 feet above the ground.

A welder repairs worn auger bits that were used to drill holes for support pipes required to install above-ground sections of the pipeline. More than 70,000 of these two-foot-diameter holes were drilled.

An above-ground section of pipe is lowered onto a "shoe" 75 miles north of Valdez.

Heat pipes were lowered by crane into vertical supports for above-ground sections of the pipeline. The thermal devices reduce soil temperatures around the supports to keep permafrost soils frozen.

A three-ton "manipulator" was used to install insulation on above-ground sections of the pipe. Operated from a crane, right, the manipulator was equipped with special suction devices (hanging below the device) and hydraulic arms to hold and wrap panels of insulation.

Heating machines were used to heat sections of pipe prior to the application of a protective tape coating.

A protective tape coating is wrapped about 48-inch pipe. This wrap was applied to buried sections of the line to protect the pipe from corrosion.

Twin ribbons of zinc wire are laid into a ditch alongside a section of the pipeline to help prevent electro-chemical corrosion of the pipe. Called anodes, the diamond-shaped wire, 1/2-inch in diameter, is connected to the pipe at 500- and 1,000-foot intervals, completing an electrical circuit caused by chemical reaction between the soil and the pipe.

The pipe received various wrappings before it was buried. This section was laid down 25 miles north of Valdez, beneath the state highway and the Tiekel River, which runs alongside. The wrappings on this section included, from left: thermal insulation material for the road crossing; pipe casing and a casing vent pipe, also required for the road crossing; felt wrapping; protective tape coating; more felt, and finally, wood slatting, which is used in river plains where concrete anchors are placed atop the buried pipe.

A panel of insulation is wrapped around an above-ground section of the pipeline. The insulation consists of 3¾-inch-thick fibrous glass, bonded to 28 gauge galvanized steel sheeting. The insulation is precut with longitudinal grooves to insure a tight fit around the pipe.

Laying pipe near the Chena River.

The route crosses the Denali Fault 200 miles north of Valdez. Concrete support beams were erected to permit substantial movement of the pipeline, which was installed in standard above-ground shoe assemblies resting on top of the beams. The pipeline in this area is designed to withstand a fault displacement of up to 20 feet horizontally and 5 feet vertically.

Expansion joints were installed between the panels of fibrous glass insulation to accommodate normal expansion and contraction of the pipeline. The joints consist of a fibrous glass filler and a flexible, weathertight bellows riveted to the insulation.

About 120 miles south of Fairbanks a valve is readied for installation. There are 151 block valves and check valves on the 800-mile-long pipeline.

A 1,500-gallon capacity hydroseeder spreads a mixture of water, grass seed and mulch over a work area near Pump Station 12. The revegetation operation is an important part of an erosion control program on the pipeline project.

A section of 48-inch diameter pipe is temporarily capped during hydrostatic testing. The piping protruding from the line is used to discharge water. Valves on the smaller pipes are shut during the test, conducted to ensure integrity of the welds and of the pipe itself.

An airplane patrols a section of the pipeline near Fairbanks. One of the methods used to maintain security, the air patrol provided information about unauthorized vehicles, damage to pipeline structures, erosion and other problems.

A Cessna Ag-wagon (C-188) begins its first pass over a 21-acre soil disposal site to be revegetated near the pipeline. The "feed" and "seed" plane makes its passes 35 to 50 feet above the drop area at a speed of 90 miles per hour. More than 11,000 pounds of fertilizer and 1,000 pounds of seed were used to revegetate this patch.

A modified FLIR (Forward Looking Infra Red) system monitors the operation of more than 122,000 heat pipes installed on the pipeline. Housed in a Bell Jet Ranger helicopter, an infra-red lens and videotape monitoring system determine which pipes need maintenance. These heat pipes reduce soil temperatures, minimizing the danger of thawing and consequent loss of soil stability.

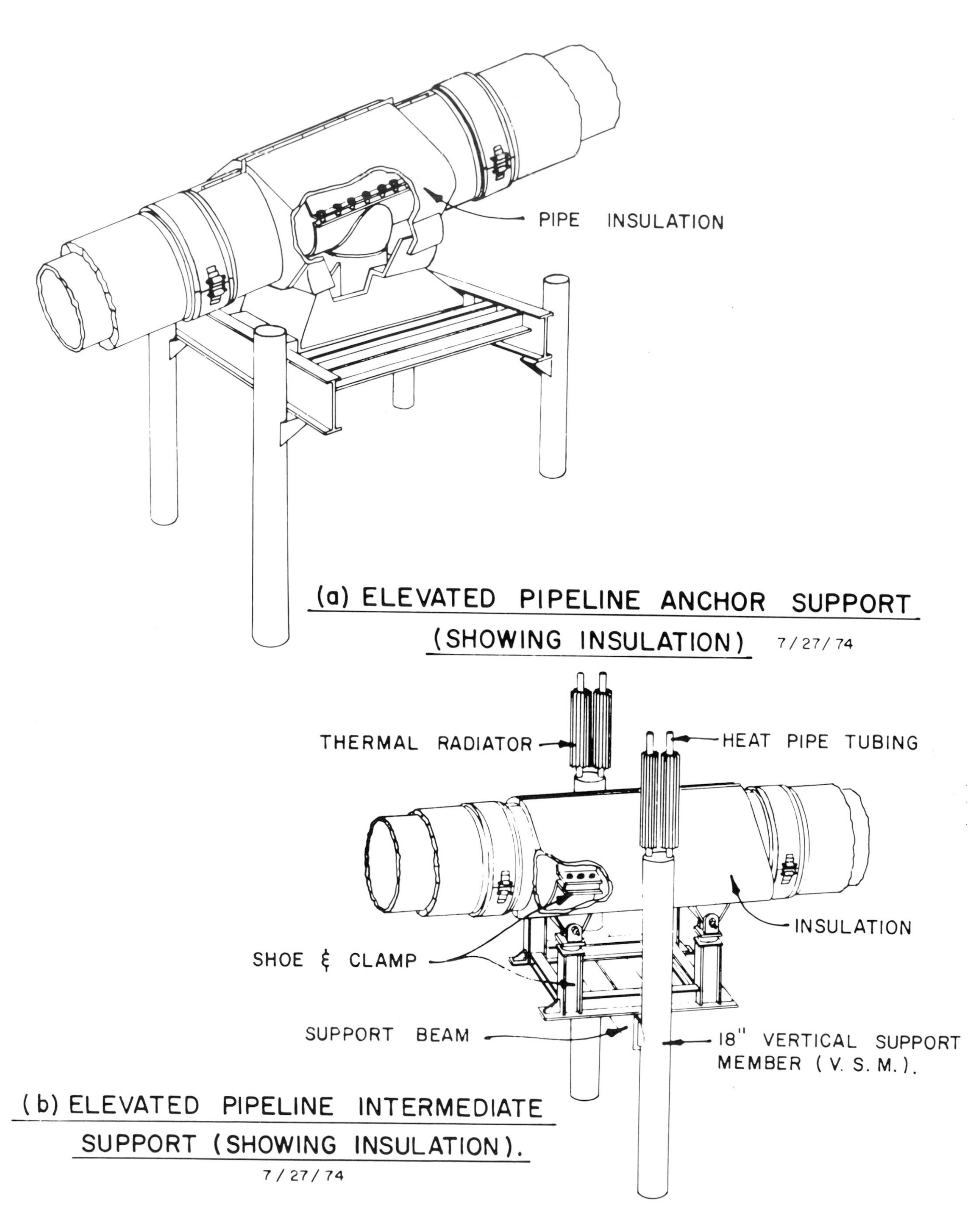
PIPE INSULATION
(a) ELEVATED PIPELINE ANCHOR SUPPORT
(SHOWING INSULATION) 7/27/74
THERMAL RADIATOR
HEAT PIPE TUBING
INSULATION
SHOE & CLAMP
SUPPORT BEAM
18" VERTICAL SUPPORT
MEMBER (V. S. M.).
(b) ELEVATED PIPELINE INTERMEDIATE
SUPPORT (SHOWING INSULATION).
7/27/74

FIGURE 4

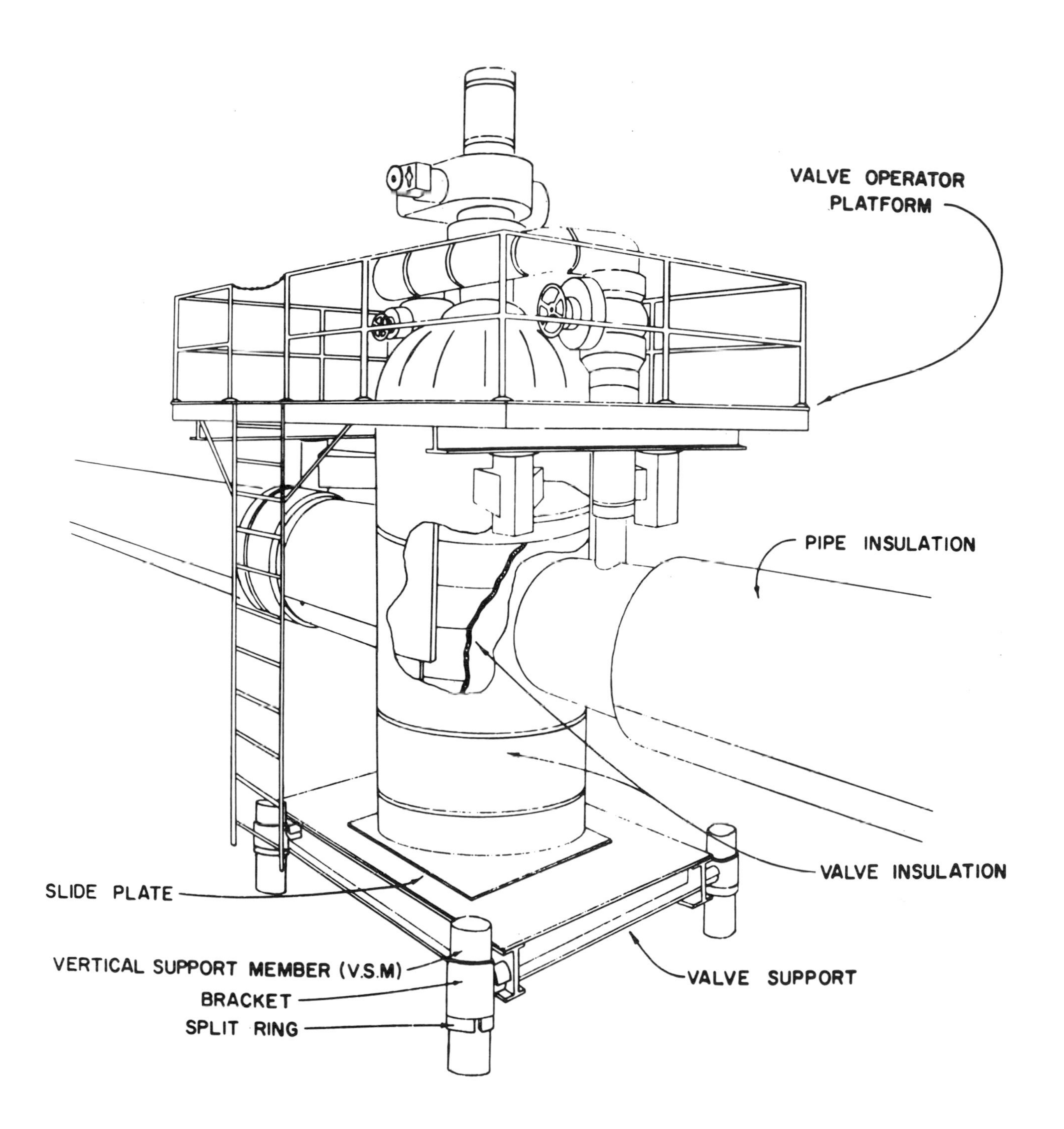

ELEVATED PIPELINE BLOCK VALVE (SHOWING INSULATION)

7/27/74

FIGURE 3

Conventional Burial

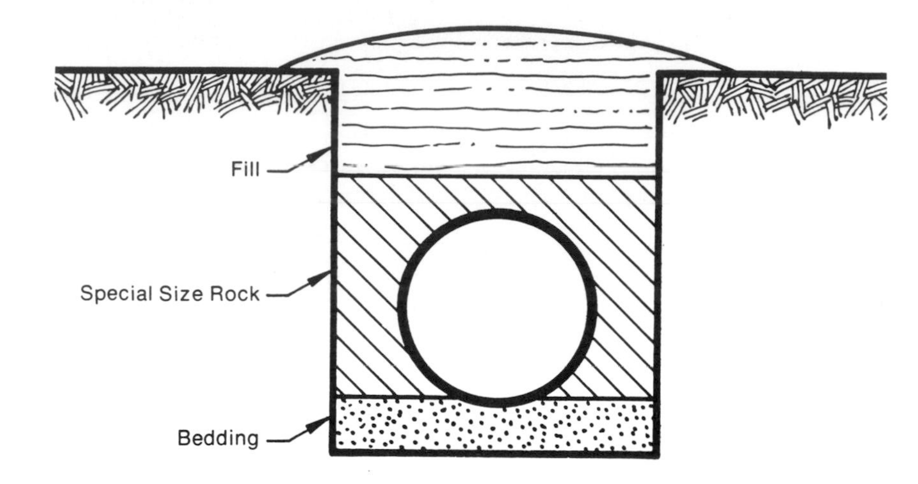

Special Burial

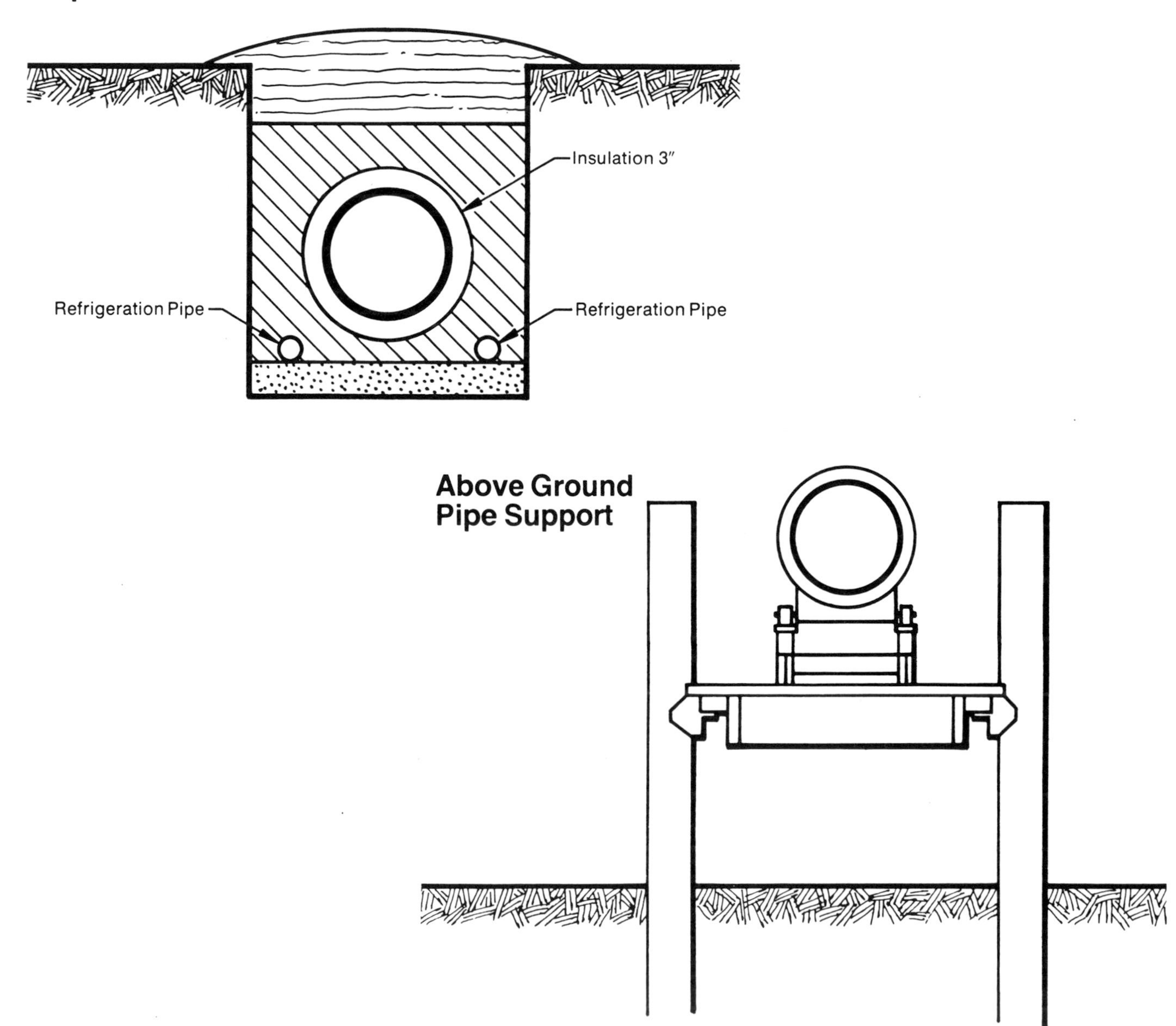

River Crossings

A river crossing in the vast, barren expanse of interior Alaska.

At 25 degrees below zero a dragline breaks ice on a water-filled ditch at the Sgavanirktok River crossing, near Happy Valley camp. Concrete-coated pipe thus slides into the water without tearing off the floats used to buoy the pipe just enough to facilitate its movement along the bottom of the 20 foot deep ditch. The pipe weighs one ton per foot with its concrete coating, while in the water, with floats, it weighs 15 pounds per foot.

The South Fork of the Koyukuk River, about 260 miles south of Prudhoe Bay, is crossed on a plate girder bridge.

He knows its cold, but how deep is it? The surveyor, supported in a clam shovel measures the depth of the Middle Fork of the Koyukuk River, about 125 miles north of the Yukon River. The pipeline was buried beneath this stream in late March 1976.

The pipeline crosses the Tanana River, about 80 miles south of Fairbanks, on a cable suspension bridge.

Mainline pipe for the Tanana River crossing was welded joint-by-joint on either side of the river and pulled to the center where it was tied in to form a 1,200 foot continuous length.

The Gulkana River bridge for the pipeline was constructed of materials already available to the project. These included surplus 48-inch-diameter mainline pipe and 18-inch-diameter vertical support members (VSMs). The large diameter pipe and the VSMs form the backbone of the bridge's substructure, its foundation piers on either side of the river.

The pipeline crosses the Gulkana River about 145 miles north of Valdez, on a tied-arch bridge.

A cable suspension bridge is used to carry the pipeline across the Tazlina River, about 113 miles north of Valdez.

Close-up view of the Tazlina River crossing.

The pipeline was buried across the Salcha River, about 35 miles south of Fairbanks. The pipe was weighted with concrete and buried at a depth great enough to ensure the river flow does not uncover and erode the pipe.

Elevations of pipeline bridges in relative scale

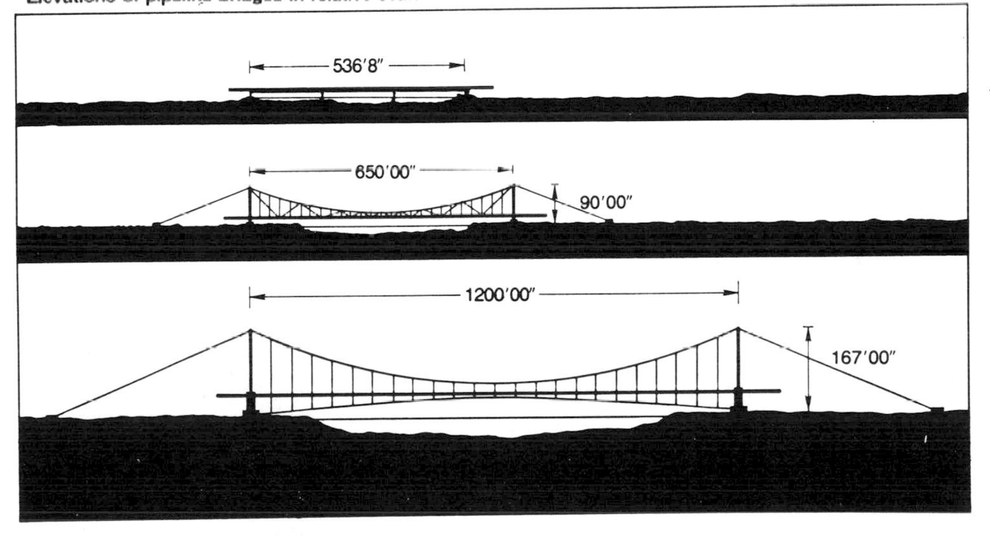

Tazlina Bridge support tower

Tanana Bridge support tower

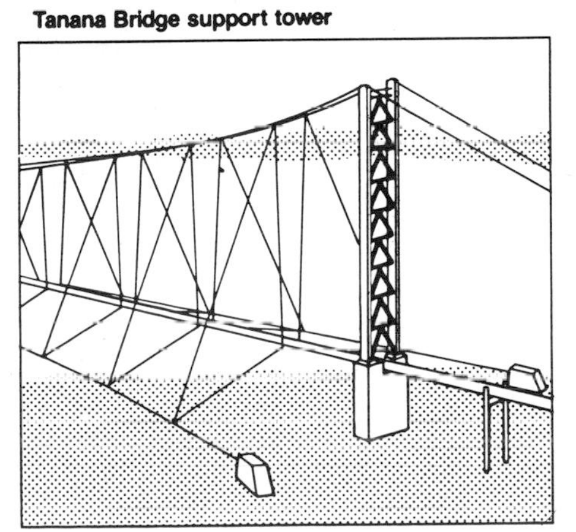

Welding

A worker buffs a weld on a section of pipe at a double jointing operation in Valdez. More than 41,000 pairs of 40-foot joints of mainline pipe were welded into 80-foot sections at plants in Valdez and Fairbanks between late 1974 and August 1975.

Welders join sections of pipe for the pipeline in the Chugach Mountains, just north of Valdez.

Welders begin one of the weld passes on a joint of 48-inch diameter pipe.

A joint of 48-inch diameter pipe is prepared for welding.

Welders work inside winter shelter.

Welders join sections of pipe.

Technicians attach a radioisotope canister to a section of pipe to re-X-ray a portion of a weld. The section of pipeline, which had been buried, was excavated to permit the remedial work, part of an intensive program to verify the integrity of all welds on the pipeline.

"Get all you can before it's taxed too" reads the message on this welder's hood. The philosopher is welding a section of pipe. The asbestos covering the pipe is to protect the pipe coating.

Welders put on one of the "fill passes" connecting joints of 48-inch diameter pipe. Each connection required an initial "root pass" and "hot pass," four "fill passes" and a final "cap pass" to complete the weld. The completed weld was then X-rayed to assure its integrity.

Ten side-boom tractors support a 250-foot section of 48-inch mainline pipe while welders in the welding hut at left connect it to another 4,250 feet of concrete-coated pipe. The concrete weights the pipe to keep it from popping to the surface of the Sagavanirktok River flood plain area.

Working inside tent structures that block the wind, welders join sections of the pipe. After the pipe was welded, it was placed in shoe assemblies and lifted atop cross supports that were placed between the vertical piles. This scene was just south of Pump Station 1 at Prudhoe Bay.

Two types of special welding cabs were used in Section 2 of the pipeline project to help welders do their jobs in adverse weather. One type, shown here, consisted of an upper section of aluminum and a lower section of canvas skirting. These units were designed for the greater mobility required in making the first two weld passes. Another unit, made completely of aluminum, was used on the "firing line," where the final passes were made to complete a weld.

An above-ground section of pipe is welded together beneath a row of protective aluminum cabs. Each unit had lighting, heating and ventilation systems to enable welders to work during extreme weather conditions.

Ten feet long and nearly 48 inches in diameter, Snoopie, an internal pipe welding and inspection tool, travels through the pipe on four rubber tracks. The vehicle is equipped with its own power lighting, exhaust and welding systems.

A radiographic crew inserts an automatic X-ray crawler inside a section of pipeline. The crawler consists of two sections. Being inserted in the pipe is the gas engine tractor which pulls the X-ray equipment, foreground, through the pipe. The X-ray crawler stops automatically at each girth weld. When the crawler stops, a pre-set exposure timer is actuated and the proper exposure is made.

CRC Internal Welding System.

R.G. Behne, Columbus, Texas, a 6-year pipeline welder helps complete final weld, Oct. 26, 1976.

About 250 craft, contractor and management personnel gathered Monday, May 31, 1977, to witness completion of the last weld made by a construction crew for the trans Alaska pipeline. More than 100,000 field and machine welds were required for the 800-mile pipeline.

■ Atigun Pass

The pipeline winds through the Brooks Range at Atigun Pass, elevation 4,800 feet—the highest elevation on the pipeline route.

About 6,000 feet of the pipeline in the Atigun Pass area is encased in a concrete insulation box, required by soil conditions in the area.

A cage of fabricated metal tubing protected workers as they spread concrete grout in a special pipe ditch section at Atigun Pass.

A concrete slab is poured in a ditch in the Atigun Pass. The slab will be part of an insulation box through which the four-foot diameter pipeline will be installed. The insulation is required because of the ice-rich permafrost soils in the area.

Thompson Pass

A section of pipe rests atop the first "bench" at Thompson Pass. Because of the steep incline, the 80-foot joints of pipe weighing between 20,000 and 38,000 pounds had to be winched up the mountain on elevated cable systems.

Explosive charges are set for ditching down the steep slope of Thompson Pass. The pipeline was installed down this slope in October.

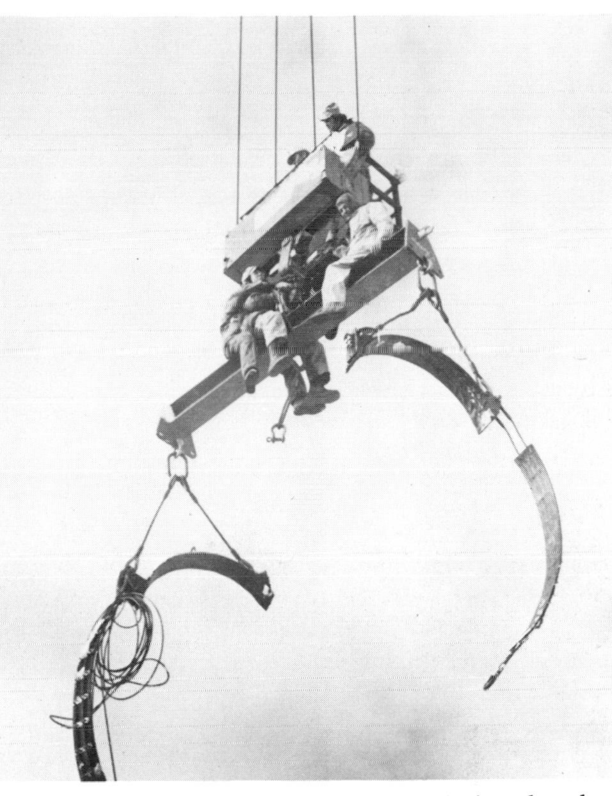

Members of a pipe gang "fly" to their work sites aboard the cable system at Thompson Pass. The huge clamps, dangling in mid-air, normally are tightly wrapped around 80-foot-long sections of mainline pipe.

Snow-covered mountains of the Chugach River loom in the background of Thompson Pass, where a cable tram was set up to haul materials and pipe up the 2,500-foot slope. At places, the pipeline rises at a 45 degree angle at Thompson Pass, near the southernmost section of the pipeline.

The cableways at Thompson Pass had working capacities of 12 tons. Here an 80-foot section of pipe, weighing about 250 pounds per linear foot, is slowly lowered into position.

Workers fight snow, cold and mountainous terrain to align two sections of pipe on the steep slopes at Thompson Pass.

Night or day, work on the 3,700-foot long Thompson Pass section of the pipeline was slow and difficult. Night shift crews used powerful portable generators airlifted to the site, to illuminate the work area.

Pipeline welders complete the "tie-in" the final weld, for the Thompson Pass section. One of the most difficult stretches of terrain, Thompson Pass, at 2,500 feet, is the third highest elevation along the pipeline's route.

Lining up for the final weld in Thompson Pass, one of the two most rugged construction areas on the pipeline, workers prepare to bring together two long completed sections of pipeline to link the line over the pass in the Chugach Mountains. Although veteran pipeliners called their finished job "the steepest 48-inch pipeline in the world," the final weld was in a relatively level area near the top of the pass. Workers also had to contend with high winds and blowing snow in the final stages of construction work in the area. The pass is about 20 miles northeast of Valdez, the terminus of the pipeline.

Last tie-in at Thompson Pass.

■ Keystone Canyon

An access road under construction winds up the south side of the Keystone Canyon, where the pipeline was placed on a bench above the canyon wall. Although the 1,400-foot-high bench in the Chugach Range is not the highest point on the route, it presented one of the more difficult engineering challenges on the project.

Pipe is buried atop the east wall of Keystone Canyon.

The pipeline is buried atop the east wall of Keystone Canyon, a three-mile long stretch in the Chugach Mountain Range.

A section of the pipeline is lowered into a ditch in a steep slope near the Keystone Canyon.

Pipe installed down the south side of the canyon.

Burial atop the east wall of Keystone Canyon.

Sideboom tractors carry the 80-foot section of pipe along the Lowe River on the south of Keystone Canyon, about 15 miles north of Valdez.

A bulldozer, modified to carry sections of 4-foot-diameter pipe, crawls up a narrow road to the top of Keystone Canyon. The three-mile canyon was one of the more challenging engineering and construction areas along the 800-mile-long pipeline route.

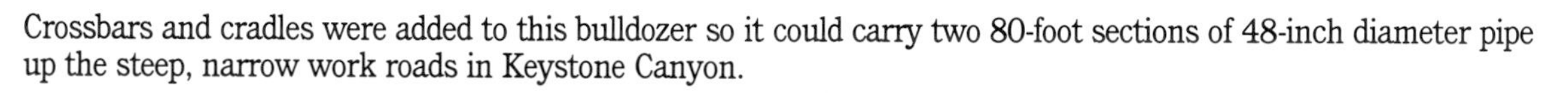

Crossbars and cradles were added to this bulldozer so it could carry two 80-foot sections of 48-inch diameter pipe up the steep, narrow work roads in Keystone Canyon.

■ Communications

Access to remote communications sites was by helicopter. This site, located about 40 miles north of Valdez, is part of a permanent microwave communications system that monitors the operation of the pipeline.

This 90-foot-high microwave communications tower, about 90 miles north of Valdez, is one of 41 stations constructed along the pipeline. The tower, the buildings at left containing electronic equipment, and the fuel storage tanks comprise each communication location.

Earth Station at Prudhoe Bay.

Communications are provided by microwave stations such as this one near Delta, Alaska. A total of 41 microwave stations, with a satellite system as backup, make up the backbone communications system for the pipeline. The system was installed by RCA Alaska Communications, Inc.

Animals

This young bull moose was camped under the pipeline just north of the Brooks Range.

Design specifications for the pipeline called for frequent animal crossings. At this site about 100 miles south of Prudhoe Bay, a short section of the above-ground pipeline was buried, allowing animals an opportunity to cross over it.

People

Women were part of the construction scene along the trans Alaska pipeline route. Pat Wilson of Talkeetna worked on a survey crew at the Marine Terminal site at Valdez. The terminal is the site of crude oil holding tanks, which receive oil from the fields almost 800 miles north at Prudhoe Bay.

5. Pump Stations

Text courtesy of Alyeska Pipeline Service Company.

CRUDE OIL IS moved down the pipeline from Prudhoe Bay to the marine terminal at Valdez by a series of 10 operating pump stations.

Eight of the stations are equipped with three mainline pumps each. Stations 2 and 7 each have two mainline pumps.

Station 5 has no mainline pumps and is not considered an operating station. It is needed for pressure relief, however, and is equipped with pressure relief facilities. The crew at this station is responsible for the operation and maintenance of more pipeline valves than any other station and also provides pipeline maintenance and emergency response capability over an extensive distance between the nearest facilities north and south of Station 5.

The Arctic weather conditions necessitated some unusual construction methods. All equipment and most of the station piping are in insulated, windowless buildings connected by enclosed hallways. Temperatures at most of the stations are designed to withstand temperatures as low as 60 degrees below zero Fahrenheit and winds up to 100 miles per hour.

The remote station sites provide living accommodations for the crew and all stations are equipped to meet all life support requirements of the crews including heat, water, power generation, sewage and waste disposal, safety, fire detection and extinguishing systems.

The facilities at pump stations include a control building, main pump building, shop and warehouse building, flammable liquid storage building, crude oil relief tanks for pipeline pressure relief when required, turbine fuel and gasoline tanks, living quarters and a water tank. Tank dikes, drainage and access roads are also included.

The shop and warehouse building includes space for the garage, raw water treatment system, sewage treatment plant, waste oil disposal system, lifeline generator system, air compressor, a fire water pump, heaters and storage.

The four-room control building houses station control panels, emergency batteries, communications systems and electrical power generation equipment.

In the booster pump building are pumps employed to transfer crude oil from relief tanks to mainline pumps.

Located within the main pump building are gas turbine-driven mainline pumps. Most of the pumps can move up to about 25,000 gallons of oil per minute, which is more than 850,000 barrels per day. The pumps at Stations 2 and 7 are provided in a different configuration, which permits operation at half as much pressure as pumps at the other stations, but at twice the volume.

Hot gas from an aircraft-type jet engine drives a turbine wheel to power a mainline pump. These gas turbines produce 18,200 horsepower.

North of the Brooks Range, turbines are fueled with natural gas from the oil fields, with turbine fuel available for emergency use.

Three southern stations have topping plants which produce turbine fuel for other stations and the Marine Terminal.

Crude oil relief tanks at the pump stations were built with cone roofs to withstand snow loads. All are equipped with mixers to assure uniform tank temperatures and prevent accumulation of sediment or wax on tank bottoms.

The tanks accommodate surges and drain-down oil during emergencies, maintenance and repair. A pressure-relief system automatically detects and relieves excessive pipeline pressure. When actuated, the valves divert oil from the pipeline to the tanks. The oil is reinjected into the mainline as soon as normal conditions resume.

Electrical power is generated at all stations. At Stations 8, 9 and 12, power also is purchased from commercial power sources. All stations have a back-up, lifeline power generation system.

A loss of electrical power at a station does not affect operation of mainline pumps because mainline turbines generate their own electrical power. Critical supervisory and communications equipment continue in operation using batteries which can sustain operation for about eight hours.

Pump station buildings and walkways, which house equipment and in which station personnel work, are heated and provided with fresh air ventilation. Most areas are heated with a heat-medium circulating system in which a fluid is pumped through heat exchanger coils in a heater and then circulated by pump to heat-use locations in the station. The heaters, capable of producing 15 million BTUs an hour, can burn either natural gas or turbine fuel.

Pump station buildings are ventilated with a positive pressure fan or a blower-driven supply of outside air. In most areas, except living quarters, the air is heated by passing it over a heat exchanger. The system is designed to provide a minimum of four air changes every hour. Under normal circumstances, building temperatures can be maintained at about 65 degrees Fahrenheit even when outside temperatures have dropped to minus 60 degrees.

Fire and vapor detection systems are installed in all pump station buildings, and all buildings are protected by fire suppression systems.

If hydrocarbon vapors in an area or zone reach 20 percent of the lower flammable limit, the system sounds an alarm and activates high-speed ventilator fans. If fans fail to correct the situation and gas concentrations reach 70 percent of the lower flammable limit, equipment in the area automatically is shut down.

Detectors of products of combustion, flame and heat initiate alarms throughout the station and in the station's control room. The flame, heat and vapor detection systems stop all heating and ventilating fans and close all building dampers prior to actuation of suppression systems.

Halon 1301 gas is discharged manually or automatically to prevent ignition or extinguish fires. Cylinders of the extinguishing gas are located in or close to the areas they protect, with the number and size of the gas cylinders dependent on the concentration of halon required. A foam and water system provides the vapor suppression ability also needed in fighting such fires.

Pump stations utilize centrifugal pumps powered by modified 13,500 horsepower aircraft-type gas turbines. Each pump can move 25,000 gallons of oil a minute or up to 850,000 barrels a day (one barrel equals 42 gallons).

Pump Station 1

Pump Station 1, the receiving station for Prudhoe Bay and Kuparuk crude oil, has several facilities not necessary at other stations. These include a system of meters for measuring incoming crude oil, three booster pumps and two 210,000-barrel surge and relief tanks.

The station's metering system measures all oil received from the North Slope producing facilities. Oil from the producers may be moved directly into the pipeline or to surge tanks to accommodate any differences between producer output and pipeline pumping rate. The tanks are equipped with mixers and heating elements designed to keep the temperature of the oil at a minimum of 40 degrees Fahrenheit. Oil comes from the ground at as much as 180 degrees Fahrenheit, and enters the pipeline at about 145 degrees.

A booster pump delivers oil from the tanks directly to the mainline pumps. The three two-stage booster pumps are driven by 1,140-horsepower gas turbines and each is designed to pump 19,750 gallons of oil a minute.

Natural gas from the oil fields is piped to Station 1, metered, passed through a gas compression and cooling facility, and then flows through a small-diameter line to the other pump stations north of the Brooks Range where it is used to power the gas turbines. The gas reaches Pump Station 1 at between 650 and 750 pounds per square inch.

Crude oil enters the trans Alaska pipeline at Pump Station 1. The two tanks that may receive the incoming crude oil are in the background.

Pump Station 1 at Prudhoe Bay on the North Slope is the origin of the pipeline.

The origin of the pipeline, Pump Station 1 at Prudhoe Bay, shown under construction.

A turbine is lowered into place at Pump Station 1 during the construction phase. Ten operating stations were constructed along the pipeline to move the crude oil south over its 800-mile route.

Chicago Bridge and Iron built the two 210,000 barrel capacity tanks at Prudhoe and 18 510,000-barrel capacity holding tanks at the Marine Terminal at Valdez.

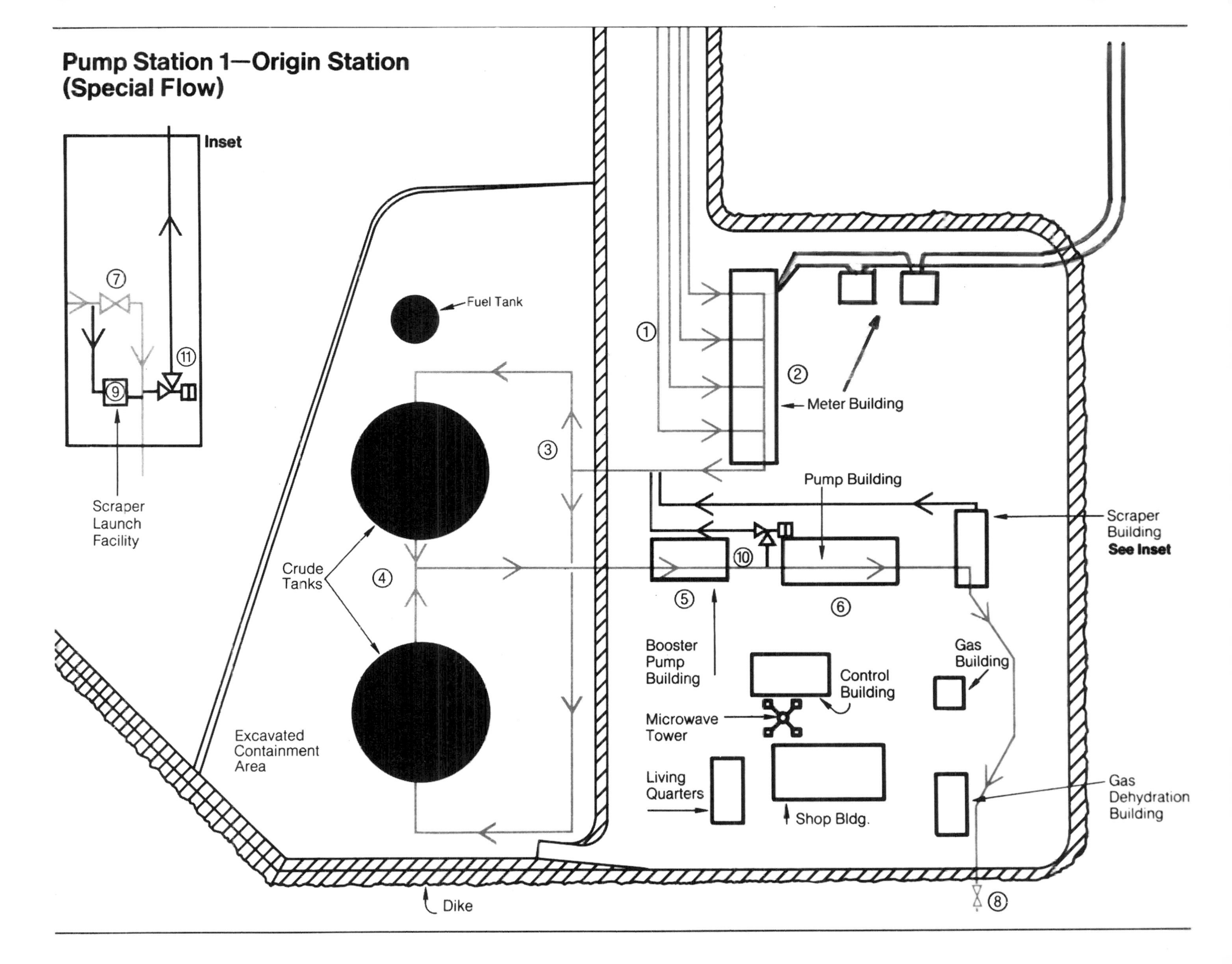
Pump Station 1—Origin Station
(Special Flow)
Inset
Scraper
Launch
Facility
Fuel Tank
Crude
Tanks
Excavated
Containment
Area
Dike
Meter Building
Pump Building
Scraper
Building
See Inset
Booster
Pump
Building
Control
Building
Microwave
Tower
Gas
Building
Living
Quarters
Shop Bldg.
Gas
Dehydration
Building
1
2
3
4
5
6
7
8
9
10
11

Pump Station 2, 57 miles south of Prudhoe Bay, under construction. This was not one of the original pump stations, but was added in October 1979.

Pump Station 3 is located 104 miles south of Prudhoe Bay.

The 48-inch pipe installed in the pipe corridor at Pump Station 3.

Pump Station 4 is located just north of the Brooks Range, 144 miles south of Prudhoe Bay. The Atigun River is at the top of the photo.

Pump Station 5

Pump Station 5, the first station south of the Brooks Range, functions as a pressure relief station. If pipeline flow stops, oil from Atigun Pass is drained into the tank at Pump Station 5 to avoid a buildup of pressure in the line. The 150,000-barrel tank is equipped with a heater and mixer.

During normal operation, incoming pipeline crude oil passes through the manifold building and flow meters and then back into the 48-inch pipeline. Whenever pressure exceeds established limits, quick-acting relief valves divert oil to the crude oil relief tank. There are two high-volume relief valves in the suction (incoming) line and two in the discharge line.

Two injection pumps and a booster pump are used to inject oil from the tank back into the main pipeline. The two-stage, centrifugal injection pumps are driven by 1,100-horsepower gas turbine units. The normal injection rate totals 100,000 barrels a day. The station's single booster pump, powered by a 400-horsepower electric motor, is used to boost oil from the tank to the injection pumps.

Pump Station 5 on the downhill side of the Brooks Range, 274 miles south of Prudhoe Bay. The station, under construction here, serves as a relief or "drain down" station.

A curious bear visits construction workers at Pump Station 5.

About one mile south of the Yukon River is Pump Station 6. The Yukon River is in the background.

Manifold building at Pump Station 6.

Pump Station 7, 414 miles south of Prudhoe Bay and 40 miles south of Fairbanks, was added to the pipeline in December 1980.

Pump Station 8 is 489 miles south of Prudhoe Bay and 35 miles south of Fairbanks.

Pump Station 10 is 585 miles south of Prudhoe Bay and north of Paxon on the Richardson Highway.

The last pumping station along the pipeline route is number 12, 65 miles north of Valdez.

Topping Plants

Small refineries, or topping plants, at Pump Station 6, 8 and 10 produce turbine fuel for the gas turbines at stations south of the Brooks Range. The topping plant at Station 10 also produces naphtha. The naphtha is used to supplement the turbine fuel at Station 10, thus making additional turbine fuel available to power pumps at other stations.

Each topping unit processes crude oil from the pipeline at the rate of 12,000 barrels a day, producing about 3,300 barrels of fuel. The crude oil removed from the line is preheated and then pumped through a heater to a distillation tower at 595 degrees Fahrenheit. In the tower, heavy oil products are pumped from the bottom, cooled and returned to the pipeline. Turbine fuel is drawn from the side of the tower at 465 degrees. Lighter fractions which rise to the top of the tower are condensed, cooled and also returned to the pipeline.

Processed fuel is held in two 20,000-barrel tanks at the topping unit stations. Fuel from the topping units is delivered by truck to the other stations and the Terminal.

A section of a natural gas fuel pipeline is lowered into a ditch about 110 miles south of Prudhoe Bay. The first 40 miles of the line is 10 inches in diameter; the remaining 110 miles is of 8-inch diameter pipe. The line will power the pump stations on the trans Alaska pipeline north of the Brooks Mountain Range.

A special 90-ton crane is assisted by a 50-ton crane to set in place the flare stack for the topping unit at Pump Station 8. Topping units at stations 6, 8 and 10 provide fuel for turbines, which powers pumps at stations south of the Brooks Range. North of the range, stations are powered by natural gas from Prudhoe Bay.

The 8-inch-diameter fuel gas line that will power pipeline pump stations north of the Brooks Range lies in a ditch awaiting burial.

To the right of the above-ground pipeline is the ditch containing an 8-inch-diameter gas fuel line that will power the pump stations north of the Brooks Range.

Welding of sections of the 150-mile natural fuel gas line between Prudhoe Bay and Pump Station 4, north of the Brooks Range, took place in small, temporary shelters, seen in the background.

This large tractor-mounted chain saw with a blade 14 feet long, is used to dig the ditch for the 150-mile-long natural gas fuel pipeline for the pump stations on the pipeline north of the Brooks Range.

Pipeline "Pigs"

Special devices called "pigs" are used to clean any accumulated wax from interior pipe walls, to survey interior pipe diameter where small changes can provide advance notice of undesired changes in pipeline condition, and to detect corrosion on the inside or outside walls of the pipe. These are put into the line or removed from it at three pump stations and the Marine Terminal.

Pump Station 1 has a launching facility. Launchers and receiver traps are installed at Stations 4 and 10. The Terminal has a receiver trap only. At other stations, the moving pig is automatically passed through the manifold by a system of valves without interrupting mainline pumps.

A pig is launched at one launching station and removed at the next receiver station. The pig may then be relaunched to continue down the line or be trucked back to its launching station.

Station pig facilities are equipped with cranes, necessary valves, drain sumps, audible alarms and a signal system linked to pump station control rooms.

Stations without launchers or receivers are equipped with pig arrival and departure signals and with valves needed to pass the scraper through the pump station manifold without stopping the flow of oil.

Traps consist of a launching or receiving tube; a trap closure and tray assembly mounted on steel tracks, which removes or inserts the pig; and winch assemblies for moving the trap closure assembly and the pig in or out of the tube.

All traps are equipped with pressure interlocks which prevent them from being opened when inlet valves have not closed or when there is pressure in the trap.

Scraper pigs are made of molded urethane, and composed of a nose cone and three cups or scrapers, mounted on a central body. The entire assembly fits the curvature of the 48-inch pipe. The cups scrape deposits of paraffin and other material from the interior walls of the pipe, and these deposits either redissolve or travel in the crude oil stream to the receiving traps at Pump Stations 4 and 10 and the Terminal. Cleaning pigs are run through the line on a regular basis.

Another type of pig, called a corrosion/deformation pig, is passed through the line at least twice a year. This instrument operates electronically to probe for irregularities in the shape of the pipe. A flat "side" or oval contour in a segment of buried pipe, for instance, could indicate unusual stresses at that location, perhaps resulting from settlement of the foundation soils around the pipe.

This pig also is employed to detect corrosion. Tests have shown that the special instrument can detect and precisely locate corrosion pitting on either the inside or outside of the pipe.

An early scraper pig.
COURTESY DELTA JUNCTION CHAMBER OF COMMERCE

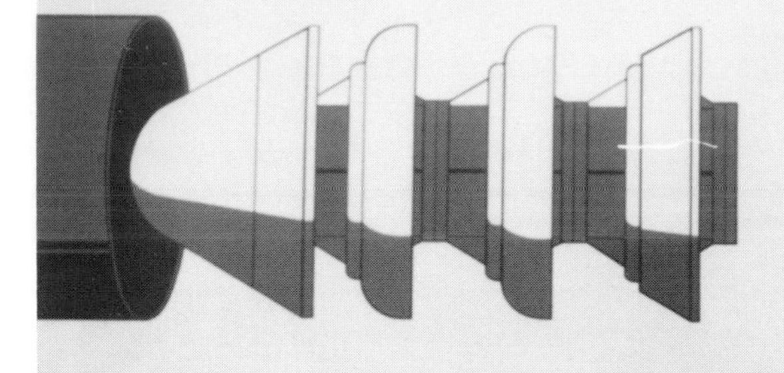

A modern scraper pig.

This Super Pig was an experimental instrumented pig that proved to be impractical due to its jointed construction.

Refrigerated Sites

Pump Stations 1, 2, 3, 5 and 6 are in permafrost areas. To maintain the stability of the permanently frozen soil, ground under most buildings and tanks is refrigerated. Coils of pipe for circulating brine are buried in gravel beneath mats of plastic foam insulation to keep the soil frozen and stable.

Mechanical refrigeration units chill a calcium chloride brine that is circulated in the subsoil piping loops at approximately 10 degrees Fahrenheit. Each unit consists of two refrigerant compressors, two air-cooled condensers, two liquid refrigerant receivers, a brine chiller and interconnecting plumbing. Temperature detectors wired to indicator alarms automatically monitor system operation.

Insulated pipe in a refrigerated ditch outside Pump Station 3.

6. Pipeline Terminal at Valdez

Text courtesy of Alyeska Pipeline Service Company.

OIL, AFTER BEING MOVED through pump stations and the pipeline from Prudhoe Bay, is loaded aboard tankers at the Marine Terminal on the south shore of Port Valdez.

The 1,000-acre site, across the bay from the city of Valdez, is on an 11-mile-long fiord in the northeast corner of Prince William Sound. It is the northernmost ice-free harbor in the United States, and offers a deep-water channel with a minimum width of approximately 3,000 feet.

Facilities at the Terminal include everything necessary for receiving oil and loading it aboard tankers. Holding facilities include 18 cone-roof tanks with a total capacity of 9,180,000 barrels of oil. Tanker loading facilities include one floating and three fixed berths which handle tankers of 16,000 to 265,000 deadweight tons at loading rates up to 110,000 barrels an hour.

Shore facilities include a vapor recovery system, Ballast Water Treatment plant, analytical laboratory, power plant, warehouses and shop buildings, meters and meter-proving equipment, water treatment and sewage systems, heating systems, oil spill contingency equipment, fire-fighting systems, fuel storage and the Operations Control Center for the entire pipeline system.

Incoming oil enters the Terminal through the last pipeline gate valve, and passes through incoming Terminal meters where it is measured. The incoming stream is checked routinely for temperature, vapor pressure, specific gravity, and other characteristics.

Other metering systems measure oil loaded into tankers. Sixteen-inch turbine meters are operated in parallel sets to measure flow, with the number of meters determined by flow rate. Each turbine meter can measure up to 25,700 barrels per hour. Meters are calibrated routinely by a meter proving system. In the prover, a "prover ball" is moved by oil flow through a carefully calibrated length of pipe. Measurement is accomplished as the device trips switches at the beginning and end of its journey through the pipe. Volume readings, reported by the meters during that period, then are adjusted to conform to the known volume of the prover loop.

Incoming facilities include a scraper, or pig, trap and pressure relief valves. Incoming pigs are received in a scraper trap in the East Metering Building, then trucked north for re-use. Scraper facilities include the trap, its valves, arrival indicators and a sump and sump pump.

Relief valves prevent incoming oil pressure from exceeding design limits. Surge or static pressure of more than 300 pounds per square inch causes these valves to open and divert oil directly to two crude relief tanks. The valves open and close automatically.

Holding tanks at the Terminal are 250 feet in diameter with a capacity of 510,000 barrels of oil each. The 18 tanks will hold about 4½ days of pipeline throughput at the 2.1 million-barrels-per-day rate.

Tanks, 62 feet 3 inches high, have a normal working product height of 58 feet 6 inches. The difference provides a 3-foot 9-inch slosh zone to allow oil movement caused by seismic forces as great as 8.5 on the Richter Scale. Each tank has three mixers to prevent stratification of the oil, level and temperature gauging instruments, and instruments designed to control the tank vapor space pressure and the positions of all valves.

Tanks are arranged in pairs having the same elevation. Each pair is surrounded by a dike which can hold 110 per cent of the oil in both tanks plus water that might be impounded in the area.

Each tank receives inert gas from the vapor recovery system through a 16-inch connection on the tank roof. The space between oil and the roof of each tank always contains inert gas rather than air. Any excess inert gas is collected via a 30-inch line on the roof and is used to blanket other tanks. Displaced hydrocarbon vapors are used as fuel to heat steam boilers at the Terminal's power plant, thus reducing the consumption of diesel fuel at the Terminal.

Tanks are equipped with 14 automatic pressure/vacuum vents which operate only in event of failure of the vapor recovery system.

The tank farm fire control system consists of subsur-

The site of the tanker terminal under construction is across from the city of Valdez, foreground.

face foam injection systems inside the tanks and a water system which can provide cooling water on tank exteriors. Crude transfer pumps can move oil from one tank to another anywhere in the tank system.

Berths at the Terminal have been numbered 1, 3, 4 and 5. Berth 1 is a floating berth; the others are conventional fixed berths. The floating berth handles tankers of 16,000 to 120,000 deadweight tons. Berth 3 handles tankers up to 250,000 DWT, and Berths 4 and 5 handle 265,000 DWT tankers. Berth 1 also can receive fuel from tankers for the Terminal's fuel supply.

The floating berth, located in an area where a steeply sloped sea bottom made conventional piling construction impractical, is held in position offshore by struts anchored to bedrock on shore. The floating structure is supported by buoys, each approximately 45 feet tall and 22 feet in diameter.

Shore struts support roadways, walkways, and oil, ballast water and fuel oil pipelines.

The fixed berths are mounted on steel structures of jacketed pile anchored to bedrock. Loading structures on these berths are approximately 122 feet long by 46 feet wide. All of the berths have mooring dolphins, or stanchions, of a conventional pile design to hold the ships' mooring lines. Mooring and breasting dolphins are all equipped with quick-release hooks.

A structure on each berth houses the berth operator's control room. From the control room, the operator has a clear view of the berth and the tanker deck with the aid of closed circuit television monitors.

Oil is gravity-fed to tankers at each berth through four hydraulically controlled metal loading arms. Ship ballast also is pumped out through these arms. The four 12-inch arms at Berth 1 handle 80,000 barrels of oil an hour. The four 6-inch arms on the fixed berths have a capacity of 110,000 barrels an hour.

Flow control and quick-shutoff valves are provided at all berths to prevent excessive surge pressures and to control flow rates during loading. Work areas on the berths are surrounded by oil-tight curbs so that any spillage which might occur can be collected and processed through the ballast water treatment facility.

The berths can be operated simultaneously and independently whether tankers are discharging ballast or loading oil.

Valdez before the 1964 earthquake. The town was almost destroyed and had to be moved a few miles away to its present site. COURTESY U.S. CORPS OF ENGINEERS

Valdez pre-construction.

The Terminal was built on the site of old Fort Liscum. The fort was established in 1900 by the U.S. Army. It was deactivated in 1922.

What might appear to some to be an early winter bean garden, is actually part of a site preparation operation at the Terminal site. The workers are setting dynamite in each of the holes where the sticks are protruding. An estimated 7 million cubic yards of material had to be moved for the facility that delivers Prudhoe Bay oil to ocean tankers.

Work is under way for site preparation of the West Tank Farm of the Marine Terminal. Four half-million barrel capacity storage tanks were built in the West Tank Farm and 14 tanks were erected in the adjacent East Tank Farm.

By the end of May 1975, 21 two-story dormitories had been completed for the construction camp at the Marine Terminal at Valdez. Twenty-eight dormitories, each with a bed capacity of about 100, were constructed at the terminal.

The snow-covered peaks of the Chugach Mountain Range rise 6,600 feet above sea level at the northeast end of the runway at Valdez Municipal Airport. The 5,000-foot state landing strip was expanded to accommodate intermediate-range commercial jets.

Crude oil holding tanks are shown under construction in 1975 at the Terminal. The tanks, 62 feet high, have a capacity of 510,000 barrels. Eighteen of the tanks were constructed.

A portion of the pipeline is installed near the Terminal. By late September 1976 almost all of the pipeline had been installed, and the project was on schedule to begin operation in mid-1977.

Work continues on 250-foot-diameter crude oil holding tanks.

Sheets of steel are placed for the floor of a crude oil holding tank. Also completed here is the first course of steel shell rings for the wall of the tank.

Workers remove snow from the roof of one of the ballast water receiving tanks at the Terminal. Almost 30 feet of snow fell in Valdez the winter of 1976.

Three 43,000-water-barrel-capacity ballast receiving tanks and 18,500,000-barrel-capacity crude oil holding tanks are required to handle the trans Alaska pipeline's production. During 1975, the three ballast trucks, foreground, and ten of the 18 crude oil tanks, background, were erected.

Workers spray concrete over a wire mesh to stabilize a slope at the Terminal.

A trestle jacket is lowered into place for one of the tankers berths at the Terminal.

Floating Berth 1 was built in Japan and towed across the Pacific to Valdez where it was installed at the Marine Terminal. These photos show the immense size of the berth.

Trestle piers extend into Port Valdez for Berth No. 4 at the Terminal. Ten trestles, joined by a 12-foot wide roadway, will extend this tanker dock 1,200 feet out into the port. Four berths were constructed at the Terminal.

Barge-mounted cranes and derricks set trestle jackets for Berth 5 at the Terminal. Fluor Engineers and Constructors, Inc., built the Terminal for the Alyeska Pipeline Service Company.

Lifting boiler stack into place.

Incinerators for the vapor recovery system at the Terminal near completion. The incinerators are used to burn off excess gases in the recovery system, which disposes of oil vapors from the crude oil holding tanks.

Power plant and vapor recovery area.

Laboratory technicians perform final chemical analyses of treated tanker ballast water before it is discharged into Valdez harbor.

Oily ballast water from tankers docking at the Terminal is treated in a process involving primary separation; chemical coagulation and dissolved-air floatation; and biological treatment. Here, an operator inspects the flow of ballast water in the dissolved air floatation basins.

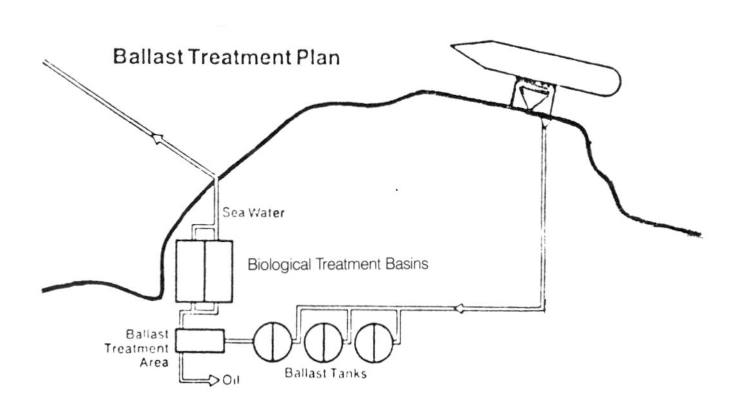

Oily ballast water carried aboard empty tankers arriving at the Terminal is pumped ashore. The water is treated and can be discharged back into the sea at less than eight parts of oil per million, on the average. The recovered oil is pumped into the Terminal's crude oil holding tanks.

Expansion loops on a pair of large diameter pipes snake upward from beneath the pier of a Terminal berth to loading arms, which are used to move crude oil into tankers. The pipe in the foreground carries crude oil flowing by gravity from storage tanks, and the other moves oily ballast water from tankers to a ballast water treatment facilty at the terminal.

Shimmering lights on Port Valdez reflect the constant activity at the 1,000-acre terminal.

The *ARCO Juneau* sails from the terminal of the Alaska pipeline with a cargo of 825,000 barrels of crude oil from Prudhoe Bay.

Oil tankers traffic through Valdez Arm is monitored by the U.S. Coast Guard. Under a detailed tanker traffic monitoring system, north and south traffic lanes are three-quarters of a mile wide, with a separation zone one mile wide between lanes.

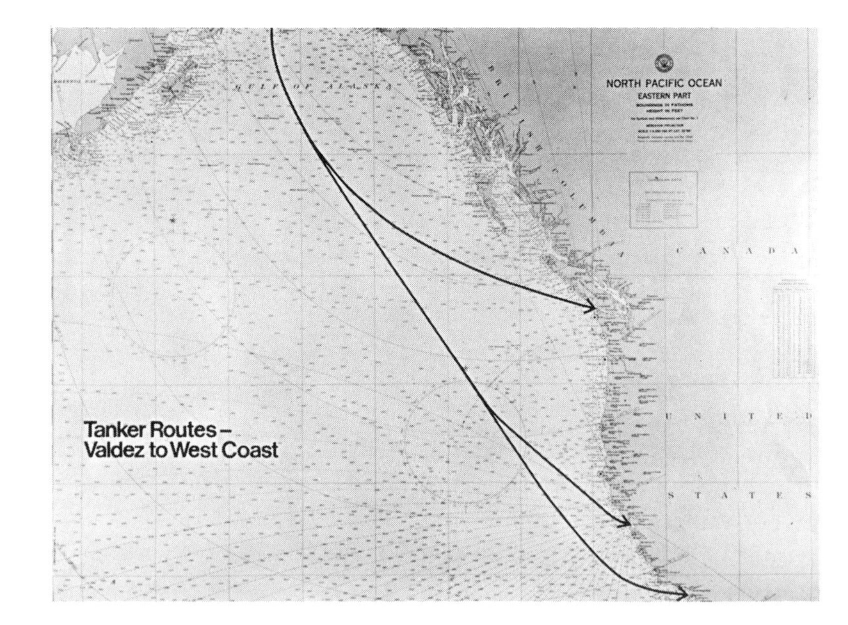

Tanker Routes – Valdez to West Coast

Power Plant

The Terminal power generation facility provides electric power to the Terminal and is a source of steam for various pump drivers and heating systems.

The Terminal does not use commercial electric power. Besides its main power generation facility, the Terminal has two back-up power systems.

The primary power plant consists of three boilers which generate steam, and three condensing turbo generators, each with a capacity of 12.5 megawatts. The systems can satisfy all Terminal power requirements.

Water for the boilers is pumped from a nearby water source into a water treatment plant where it is filtered and demineralized before being deaerated and fed into the boilers.

The "lifeline" power system, consisting of two emergency generators located in a building adjacent to the power plant and the emergency response building, is intended for use in the event of a power failure at the Terminal. The units are 12-cylinder diesel-driven generators with associated control equipment.

Four uninterruptable power systems provide a third level of power for essential control equipment and instruments throughout the Terminal. The systems consist of batteries, chargers and distribution panels along with rectifiers, inverters, necessary transformers and switches.

Vapor Recovery System

The Terminal's vapor recovery facility consists of two main systems. The first provides inert gas blanketing for all crude oil holding tanks. The second system collects the inert gas and oil vapors from the tanks when they are being filled or when atmospheric conditions cause an increase in tank pressure.

Flue gasses from power plant boilers located next to the vapor recovery facility are used as the inert gas. The flue gas—approximately 70 percent nitrogen, 18 percent carbon dioxide, 9 percent water vapor, 3 percent oxygen and minute quantities of sulfur dioxide—is cooled and then "scrubbed" to remove the corrosive sulfur dioxide, and then sent to gas compressors and, ultimately, the tanks.

The sulfur dioxide is removed in water scrubbers which eliminate 90 percent of the sulfur dioxide gas as it flows upward through a spray and over trays of scrubbing liquid. Any free water in the gas is separated in a suction knockout drum equipped with a demister screen.

The cooled and processed inert gas then is compressed and discharged to oil tanks as required.

When tanks are being filled or when atmospheric pressure decreases, tank vent valves open and allow gas—generally a mixture of hydrocarbon vapors and inert gas—to flow into the low-pressure collection system. After moisture has been removed and the gas has been compressed, it is returned to tanks as blanketing gas.

Excess hydrocarbon vapors are used to fire the power plant's steam boilers, or are burned in one of the Terminal's three thermal oxidizers.

Ballast Water Treatment

Oily ballast water carried by tankers and all other oily waters collected in the Terminal area are processed through the ballast water treatment system before being discharged into Port Valdez.

All tankers arriving at Valdez carry water ballast for stability while traveling without a cargo of oil. Some of the vessels carry only "clean" ballast, i.e., water held in segregated tanks. Others carry at least part of their ballast in tanks also used for oil when the tankers are traveling with cargo. This becomes oily ballast, because of mixing with the residue of oil previously carried in the cargo tanks.

Oily ballast water, which must be removed before tankers can take oil, is pumped to the treatment system. Tanker operators arriving at Valdez must certify that no oily ballast has been discharged at sea.

The Terminal treatment system consists of gravity separation tanks; chemically aided dissolved air flotation treatment chambers, and a biological treatment process.

Ballast water is pumped first into gravity separation tanks where oil is allowed to rise to the surface. Two floating-boom skimmers remove oil from the surface and the remaining ballast water is removed from the bottom of the tanks for secondary treatment.

In the second step, ballast water is piped to dissolved-air flotation treatment chambers where coagulant chemicals are mixed with the ballast water to cause suspended and colloidal materials to form particles referred to as "floc."

The water is aerated, and the bubbles cling to the floc, carrying it with the suspended oil and solids to the surface where it is skimmed for removal. The remaining water then flows to the biological treatment ponds, where residual crude oil components are reduced to extremely low concentrations. The water is then discharged into Port Valdez at depths of 200 feet or more, 700 feet offshore.

The recovered oil is blended into the crude oil stream for eventual loading in tankers.

Fire Protection System

Water, foam, halon and dry chemical are used in the Terminal's extensive fire protection system which includes three fire trucks and other firefighting equipment. In event of a fire, an on-scene commander directs efforts of a fire response team of Terminal employees.

Seawater taken from Port Valdez by automatic-start diesel-powered pumps is used in the looped firewater system which has a 10,000-gallons-per-minute capacity. If fire should occur inside a tank, foam would be injected below the surface of the stored oil, bubbling through the flammable liquid to form a blanket on the surface which suppresses vapor and extinguishes the fire.

Each berth has its own firefighting facilities consisting of firewater pump and foam systems. When a fire alarm is tripped, the flow of oil is stopped and the foam system blankets the loading area with foam. Each berth is also connected to the onshore system as supplementary protection.

Offshore are three tugs, equipped to deliver cooling water and foam, to supplement fixed firefighting systems on the berths. If fire occurred aboard a tanker, the tugs would supplement the onboard firefighting systems.

A halon extinguishing system is used in the Operations Control Center. The ballast water treatment building is protected by a firewater foam system.

Operations

Although the pipeline consists of separate elements with different functions, the elements must be viewed as one system. Pump stations, pipeline and Terminal operate together to move crude oil in an orderly, controlled manner to tankers at Valdez.

Control Center

This 800-mile long system is controlled from the Operations Control Center at Valdez, where Pipeline and Terminal Controllers monitor and control almost every element.

The Pipeline Controller, aided by computers, control panels and data displays, monitors and controls the pipeline by remotely operating valves and pump station equipment.

Working at a console, the Controller can initiate actions at every operating level, such as starting and stopping mainline pumps, controling valves, isolating pump stations and starting and stopping the entire system.

The Terminal Controller likewise governs certain Terminal functions, including directing the flow of pipeline oil to specific holding tanks and from the tanks to vessels at the berths; and monitoring tank levels at the Terminal's Ballast Water Treatment facility, metering systems and tanker loading berths.

The Pipeline and Terminal Controllers also can initiate a wide range of emergency procedures in case of any problems on the line, in the pump stations, or at the Terminal.

Many operations on the line can be controlled locally. At the Terminal, tanker loading is handled by berth technicians under the control of the Terminal Controller. The Terminal Ballast Water Treatment plant is operated from its own control room, with the Terminal Controller monitoring all alarm conditions.

Each pump station, likewise, can assume local control of equipment in case of emergency or during repairs. Stations, however, may assume local control only with permission of the Pipeline Controller.

Monitoring the System

In monitoring the pipeline and pump stations, the Pipeline Controller receives a constant flow of information about conditions along the line.

Data includes specific information on alarm conditions, the leak detection system, seismic events, flow rates, pressures, temperatures, specific gravities and the operating status of pumps, valves and other equipment. The information, analyzed and stored by computers, is displayed on television-like cathode ray tubes for quick review by the Pipeline Controller.

All of the information is transmitted over the pipeline's backbone communications system of microwave stations which generally parallels the pipeline route and which links pump stations and maintenance centers with Valdez. Remote valves on the pipeline are linked by dual VHF radio channels to the main network.

The microwave system is backed up by a satellite communications system. Earth stations at Pump Stations 1, 4, 5 and at Valdez can communicate with each other via a space satellite in orbit 25,000 miles above the equator. The satellite system handles all pipeline control data in event of any break in communications in the chain of microwave stations.

Response Systems

Information transmitted by the systems can trigger a variety of responses. In a few instances, signals to the Operations Control Center can initiate automatic reactions. In most cases, however, the Pipeline Controller initiates any response.

Pressure changes, communications failures, valve closures, seismic alarms—all can trigger preprogrammed, automatic responses ranging from shutting down an individual station to shutting down the entire system.

A high-level earthquake alarm, for example, will initiate an automatic sequence of events which will shut down the line unless the alarm is acknowledged at the Operations Control Center.

The control system also is set up to handle a large number of situations on a sequential basis. After the Pipeline Controller initiates a sequence, the Control Center computer orders the series of events required to execute the command.

For instance, once the Controller initiates a "station shutdown" command for any pump station, main pumps in the station are shut down and locked out in sequence; booster and injection pumps are shut down if in operation; automatic scraper bypass controls are switched to manual mode.

To isolate a pump-house, while permitting oil to bypass a station, the sequential system shuts down main, booster and injection pumps; switches scraper operation to a manual mode; cuts off fuel and gas valves, and opens valves to permit mainline oil to bypass the pumps.

Controllers also can issue a number of specific commands to particular pieces of equipment. Booster and injection pumps at stations can be started and stopped individually. Gate valves, ranging from those controlling the flow of oil from producers to remote valves along the line, can be opened and closed on command.

In addition, detailed procedures have been established for Controllers in some of the more complex operation and emergency situations. Steps for starting the system, normal shutdowns, varying flow rates, changing pumps, operating remote gate valves, moving scraper pigs through the system and dealing with seismic events are all spelled out in detail for the Pipeline Controller.

Leak Alert

Four automatic leak alert and alarm systems are provided for the Controllers: pressure deviation, flow rate deviation, flow rate balance deviation and line volume imbalance. All leak alert alarms are processed by the Control Center computer.

The pressure deviation method can indicate large leaks near a pump station site. The alarm will sound if measured pressure at a site drops from the pre-set value.

Much more sensitive is the low volume imbalance method. Oil entering the line, oil arriving at the Terminal and any oil offtakes are measured by sensitive meters whose accuracy is proved regularly. These readings are calculated by computer, with compensations for pressure, temperature, gravity and other variables, to achieve precise balancing.

The flow rate deviation and flow rate balance deviation systems depend upon Leading Edge Flow Meters, or LEFMs, which are the result of years of research by the metering industry for measurement of fluid flows using sonic signals.

The LEFMs are installed on the discharge side only at Pump Station 1, and on the suction and discharge sides at the 11 other station sites. Flow rate deviation trips an alarm if the flow rate measured deviates from the pre-set value. The flow rate balance deviation system detects differences from the pre-set value of the flow rate between adjacent stations.

All the pre-set values are deduced from operating conditions measured a fixed time before.

The automatic leak alert systems are supported by frequent visual air and ground surveillance.

Leak Procedures

If a leak is detected and located, the Pipeline Controller immediately will alert Alyeska teams responsible for controlling and cleaning up any spills.

While one team stops or limits the leak, another task force handles logistics, manpower, communications and documentation problems. Government agencies are also notified.

Crews at any spill scene are armed with detailed information on what to do. Every mile of the line has been analyzed with a possible spill in mind so that crews can identify sensitive wildlife and fish habitats and can know precisely where to erect any needed containment structures.

Terminal Controller

The Terminal Controller maintains the same high level of control over Terminal operations that Pipeline Controllers maintain over the pipeline. The Terminal control systems presents the Controller with constantly updated information on all phases of Terminal operation. Included are data on storage tank levels; valve status; incoming oil flow; berthing information; and operations of the vapor recovery system, power plant and ballast treatment plant. In addition, Controllers receive information on the rate of oil flow into the Terminal, between tanks, and from tanks to the Terminal's four loading berths; on seismic conditions; and on marine and weather conditions. A special fire protection display on the Controller's console notifies him of fire alarms in critical Terminal areas.

The Controller also commands the Terminal's metering system, and makes certain that meters have been proved for accuracy by the Terminal prover system and that records are complete.

Information received by the Terminal Controller can trigger a variety of responses. In some instances, automatic sequences are begun. In others, the Controller may be signaled to start particular procedures.

Terminal Response Systems

Step-by-step responses are spelled out for the Terminal Controller in other operating or emergency situations. During tanker loading, the Controller monitors the loading rate of oil and oil levels in discharging tanks. At the end of a loading sequence, he closes down all valves in the delivery system.

The Controller also controls valves during the transfer of oil from one tank to another, sets up the flow of ballast water from tankers to the Terminal treatment plant and monitors the flow of ballast until the discharge cycle is complete.

In the unlikely case of an oil spill within the Terminal, the Controller takes immediate steps to limit the spill and then initiates containment efforts by alerting the Alyeska oil spill coordinator and other officials.

The Controller is charged, too, with initiating responses to any tanker spills in Prince William Sound. The oil spill coordinator and Terminal Superintendent and Controller then would activate Alyeska response teams and clean-up crews. An oil spill contingency barge, fully equipped, can be towed to the site of any oil spill emergency in the Sound.

Although Pipeline and Terminal Controllers command almost every phase of the pipeline operations, some parts of the system can be controlled locally, generally with permission of system Controllers.

Localized Controls

Pump station operators can operate local mainline pumps, valves, booster pumps, injection pumps and tank mixers. They can monitor local operating conditions from appropriate gauges, check the status of pump station equipment and adjust the limits of station operation. Pump station operators are alerted by a system of audible alarms and warning lights to all local emergency situations.

Each station is equipped with a number of automatic protective devices which can shut down a piece of equipment or the entire station. Some will isolate the pump house from the mainline and others will result in shutting down the mainline. Included are fire sensing devices; heat sensors on pumps and drivers; liquid-level detectors on tanks; pressure sensors in pipes; atmosphere sensing devices in enclosed buildings; turbine speed indicators; vibration detectors; reverse rotation monitors; and flow rate sensors.

In the Terminal, tanker loading operations at the four berths are under the local control of the Terminal Marine Manager.

The Manager and his staff control all debalarms, firefighting equipment and mooring equipment before a tanker arrives.

detail. In preparation for the arrival of a tanker, berth operators are required to go through a detailed readiness check on oil loading arms to make certain the arms will function properly in all positions. They are required to check out alarm systems on the loading arms, firefighting equipment and mooring hook releases before a tanker arrives.

In receiving ballast from a tanker, the Terminal Controller sets up pipe-routing for delivery of the ballast to the treatment plant, but then grants permission to the berth technicians to handle actual discharges from berthed tankers. Berth technicians are charged with detecting any possible spills during the pumping operation.

Similar procedures are set out for oil loading. When given permission by the Terminal Controller, berth technicians can open berth valves to start delivery of crude oil to the tankers. Interlocks in valve systems block the flow of crude oil when ballast water valves are open or when any fire or danger signals have been activated.

The Berth Technician is responsible for inspecting tankers before loading begins and, also, for reviewing loading procedures and restrictions with tanker officers.

Terminal crews are to provide containment and cleanup assistance to tankers in event of any spill at the Terminal. In event of a spill near a tanker, floating containment booms are deployed around the vessel and surface oil. Skimmer boats then recover oil within the booms and transfer it to barges or other facilities. Near shoreline, booms are deployed to protect threatened areas. Contaminated water collected in the process is discharged into the Terminal ballast treatment system.

7. Pipeline Monument

Text courtesy of Alyeska Pipeline Service Company.

The Monument

On June 20, 1977, the first North Slope crude oil began its journey from Prudhoe Bay to Valdez through the trans Alaska pipeline—and history's largest privately financed construction project was officially in operation.

The Owner Companies of the pipeline felt that this tremendous accomplishment deserved the highest possible recognition. In August of 1977, R.O. Anderson, then Chairman of the Board of Atlantic Richfield Company, commissioned sculptor Malcolm Alexander to create a fitting monument to the men and women who built the pipeline.

Malcolm Alexander accepted the challenging assignment, and came to Alaska in search of inspiration. Although most of the 70,000 people who had worked on the pipeline project had moved on, some still remained—and, through hours of conversation with them, Alexander came to discover the enterprising Alaskan spirit that his sculpture so masterfully represents.

He interviewed pump station and Terminal personnel, talked with people in communities affected by pipeline construction, took pictures, collected authentic clothing and equipment, and generally got acquainted with the country. The widely varied insights, viewpoints, and opinions that he gathered in Alaska formed the basis for his creation of the Pipeline Monument.

Thirteen feet high, resting on a seven-foot base, the Monument expresses the larger-than-life qualities demanded of the pipeline planners and builders—bravery, strength, tenacity, confidence, and imagination. Their differing social and philosophical attributes, as well as their professional skills, are represented by the sculpture's five individual figures.

From the blending of cultures suggested by the Alaska Native workman to the adventurous spirit reflected in the woman Teamster, the Monument interprets the ambitions, emotions, and ideals of all the 20th-Century pioneers who came together on this incredible project. Each figure tells its own story—and each plays a special part in the whole drama of the trans Alaska pipeline.

The Pipeline Monument, a project of the eight major pipeline companies which own the trans Alaska pipeline, was dedicated at the Marine Terminal September 20, 1980. Dignitaries from all over Alaska were in attendance, as were workers, townspeople, and officials of Alyeska Pipeline Service Company—the builder and operator of the pipeline—and of the Owner Companies.

The Sculpture

Once the concept for the Monument had been formulated, and preliminary studies completed, Malcolm Alexander created an 18-inch clay model for approval, followed by a detailed 48-inch clay working model

Putting the finishing touches to the monument.

Dedication of the monument, Sept. 20, 1980.

weighing more than half a ton. From this stage on, the task of completing the work became one of engineering rather than direct sculpture.

When the 48-inch model was completed, a plaster mold was made from it and shipped to the Parks Studio in Wilmington, Delaware, where it was used to produce a fiberglas model.

The Parks Studio was able to reproduce the 48-inch fiberglas model as a 12-foot clay model. Molds of each section of this model were prepared, and were shipped to the Tallix, Inc. Art Foundry in Peekskill, New York.

Again, a series of molds was made—first in plaster, then using a rubberlike substance. Hot wax was poured into the rubber molds; and, when the wax cooled, the rubber was gently peeled away, leaving perfect wax models of all sections of the sculpture. A tube—or "gate"—through which bronze would later be poured was inserted in each model section.

The bronze casting method used by Malcolm Alexander in the Pipeline Monument is known as the lost-wax process—an accurate and descriptive name. The wax molds are dipped in a milky silicone solution, which builds up and hardens into a shell as it dries. When the shell is about one-half-inch thick, the coated wax models are heated until the wax melts, runs out of the mold, and is "lost."

When this stage had been finished, molten bronze was poured into the molds. After cooling, the shell-like molds were chipped away, and the bronze castings were complete.

The assembly of these castings into the total piece then began. The Monument had been cast in 86 parts, all carefully numbered for correct identification; and the original 48-inch working model was constantly referred to for accuracy. The sections were welded together on both the inside and the outside, to ensure that there would be no break in the sculpture's surface. Any small holes could, in time, develop into large holes, destroying both the appearance and the structural integrity of the work; so special care was taken in the welding process.

The final stages began with "chasing," in which the entire surface of the sculpture was finished to remove the welding surfaces. Then color was added, by heating the metal and applying acids to produce the desired hue, or patina. Finally, the sculpture was thoroughly waxed for protection—and the Pipeline Monument was finished.

During the long weeks of casting and construction, the City of Peekskill was supportive, interested—and very curious. So, by special request of the Mayor, an all-day Open House was held at the Tallix Art Foundry upon completion of the Monument. The citizens of Peekskill came by the scores to admire the massive sculpture, and Alexander was presented with the keys to the city.

The entire modeling and casting operation was overseen, from beginning to end, by Malcolm Alexander. His dedication to his work guarantees that the finished Monument is a faithful reflection of the inspiration—and the artistry—of its creator.

The Sculptor

Malcolm Alexander was born in Detroit. His father, an automotive engineer, encouraged his son to prepare for a career more financially rewarding than art—so after a stint in the U.S. Marine Corps, Alexander earned a degree in psychology from the University of Texas. A war injury having cut short a promising athletic career, Alexander gravitated toward the arts, to which he'd been strongly attracted since childhood.

He began his career as a painter—but, several years ago, he put away his brushes and paints and began to sculpt in earnest. He has created his masterworks in many different areas of the United States, and in England, Spain, and Greece.

Malcolm Alexander has long been respected for his outstanding sculptural talent—a gift which is immediately apparent in his evocative renderings of heroic Americans, including athletes both famous and unknown. His "Backbone of America" and "Working America" series have justly won international acclaim, as have many of his other works.

The sensitivity and philosophy evident in all Malcolm Alexander's sculptures are the same elements that make his Pipeline Monument a meaningful tribute to the hardworking, courageous men and women who struggled against the clock, the climate, and the rugged Alaskan wilderness to build the trans Alaska pipeline.

8. End of Project

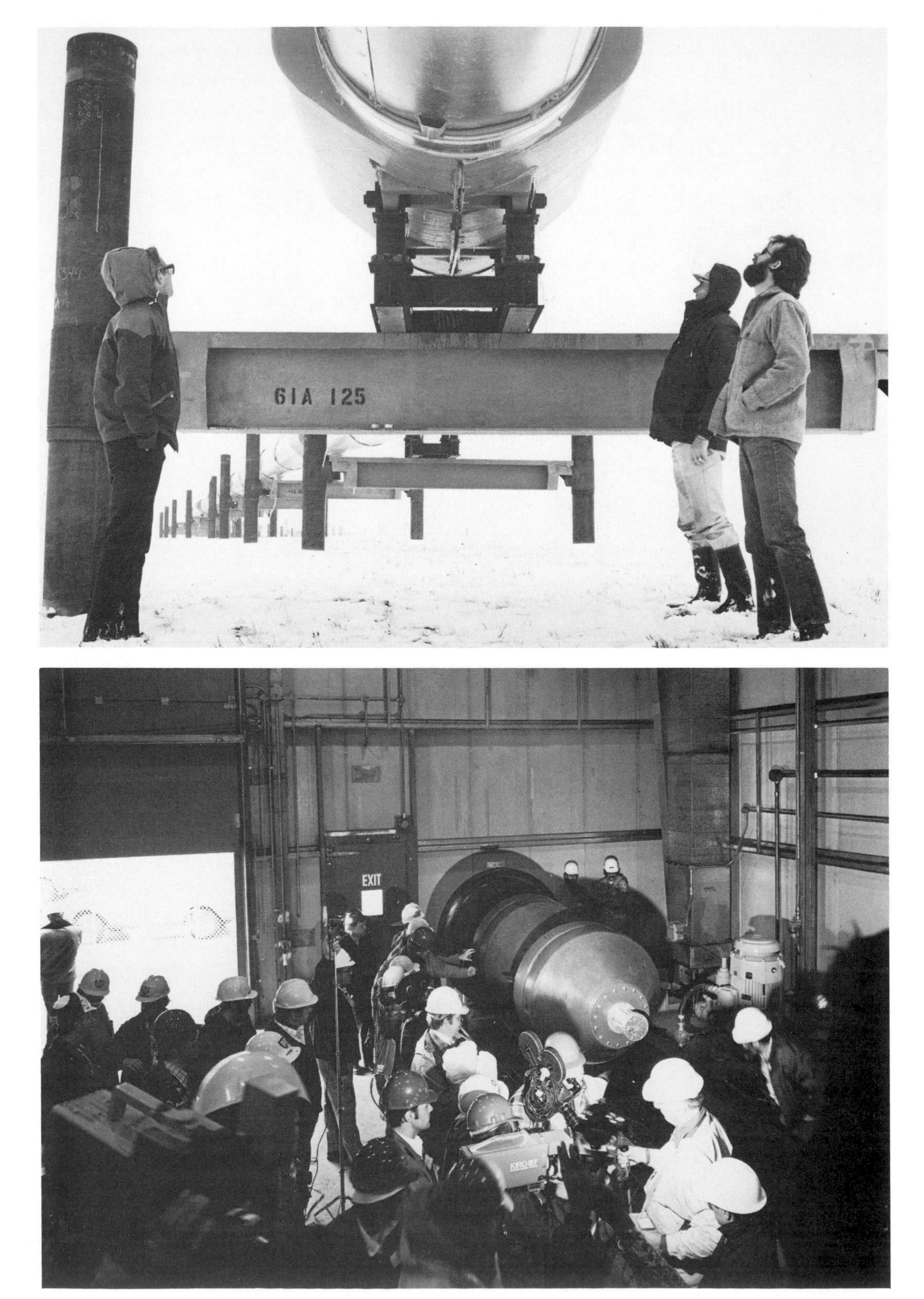

A mid-summer snowfall didn't slow the first flow of crude oil through the trans Alaska pipeline. Listening to the oil front pass overhead are, from left, Henry Mowell, Harry Robertson and Mike Jens. The oil reached the tanker terminal at the Terminal at Valdez in late July 1977.

About 60 members of the news media gathered by the pig launcher at Pump Station 1 on the moring of June 19, 1977, one day before the actual start-up of the pipeline.

Alyeska Pipeline Service Company engineer Bruce Markey tracks the location of the crude oil front in a buried section of the pipeline using an electronic device, which received signals from a transmitter attached to a metal pig. The conical-shaped pig was pushed ahead of the crude oil in the pipeline.

Equipment, facilities and materials no longer needed for construction of the pipeline were sold as surplus. Some 20,000 units of equipment originally valued at about $800 million was disposed of over a two-year period.

The Marine Terminal.

Pump Station 12 north of Valdez.

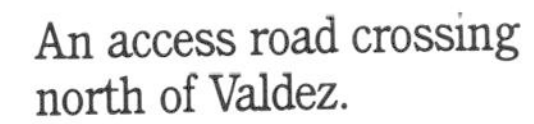

An access road crossing north of Valdez.

Plaque located next to the visitors' center.

Marine Terminal entrance and visitors' center to the right.

Loading oil at the Terminal.

Tanker in Valdez Arm.

VOL. III, NO. 50 | Alyeska Pipeline Service Company – An Equal Opportunity Employer | December 15, 1976

Mainline Pipe Installation Is Complete

All pipe installation on the line is complete, with some backfill work remaining at final tie-in locations. Two special construction areas, Atigun Pass in the Brooks Range and Thompson Pass in the Chugach Mts., were final milestones in the 1976 construction year. Severe weather conditions hampered efforts at both sites.

Atigun Pass, where pipe had to be buried in a special insulated box, due to unstable permafrost in the area, was completed on Thanksgiving Day.

One of the most spectacular, yet time-consuming operations carried out on the trans Alaska pipeline project came to an end late on Sunday, Dec. 5, as the last backfill went into the pipe ditch atop Thompson Pass.

The Radmark machine which was used to blow fill material upslope and into the ditch was shut down for the last time almost on the stroke of midnight, according to Alyeska senior project engineer Jack Cave.

Atigun Pass Work Continues Into The Night

Cave said that crews were so tired when the work ended there wasn't even a suggestion that a celebration was in order.

Hydrotesting of the newly backfilled pipe began on Monday, and was completed Tuesday afternoon, he said, pointing out that crews began removing equipment from the slope of the pass almost immediately after testing the pipe, so that the area would be secured for the winter. Clean up and dressing of the area will be completed next spring.

The importance of wrapping up the backfilling job became apparent as Cave reeled off names of high-level MK-R executives on hand to help speed work at the site. Included were G.W. Gilfillan, international vice-president of the firm, who acted as operations manager; W.H. Whitman, equipment manager acting as construction manager; Murray Swanson and Jack Granger from the MK offices in Boise, along with R. Bostwick, who was moved temporarily from duties at the Valdez Terminal. Others in the group included Ellis Mercer, Dave Fields, Jim Keene, Bill Bradfield, Don Hand, Bill McIntosh and Kiwi Callan.

Noting that snow atop the pass had drifted to heights of about 20 feet, Cave said MK-R crews constructed a special heated shelter that was placed over the ditch to keep snow from blowing in after the ditch was cleared for backfill. The 30 by 80 foot framework of pipe covered with plywood required a 90-thousand-pound capacity winch to move it up the slope. Its interior was heated by as many as six one-million-BTU heaters.

Completion of Thompson Pass concluded major construction activity in Section 1. Major activity also is complete in Section 2. In Section 3, one fifth of a mile of insulation and nearly 32 miles of hydrotest remain, and in Section 4, 7.4 miles of insulation remain. Work continues on almost 33 miles of insulation in Section 5, where 130 miles of pipe are scheduled for hydrotesting in 1977.

Work also continues on the fuel gas line from Prudhoe Bay to P.S. 4. Total project manpower was 9,282 on Dec. 5.

Innkeeper Shapes Garbage Into Art

When most of us look at garbage that's what we see—garbage. But for Alyce Reed, Atigun innkeeper, it's material for art projects, for studies of her own ingenuity and creative expression.

If Alyce's art didn't impress you, her presence would. It's not that she's agressively apparent. On the contrary, Alyce is quiet and prone to avoid conversations about her accomplishments. It's a quality in her that draws you, not any flamboyant gestures or behavior patterns.

Alyce hails from New York City where she has been home editor for Essence Magazine and a major contributor to Family Circle. She also has done such things as being assistant producer to "Hair" and working with the Alvin Aliey Dance Company.

By working on the pipeline, Alyce hopes to be able to support herself creatively with the backing of the money she accumulates north of the Yukon. "I've just begun. Ultimately I want to write but I experience with my hands. I like metal. It's inexpensive."

With tin cans Alyce has made Christmas trees, tables and other items. She also has made couches out of pipe and chains. Even Atigun has received her creative aid. On halloween she contributed eye glasses made from dixie cups—an idea she came up with for Hallmark.

Creativity flows from Alyce and infects anyone around her. All of a sudden you're thinking of all the things you've wanted to do and haven't. And if you feel inspired by Alyce, you are far

(See Page 4, Col. 1)

Completed Above-Ground Pipe Near Pump Station 4

VOL. IV, NO. 19 | Alyeska Pipeline Service Company – An Equal Opportunity Employer | May 11, 1977

Pipeline Camps Were Homes To 60,000

By VICKY STERLING

Pipeline construction camps.

In about three years, 18 of them served as homes to some 60,000 individuals.

They were the subject of much curiosity—"Say, what's it really like up there?"—and the subject of much speculation: Stories of loose living, gambling and gourmet meals were chasers for tales of isolation, of freezing to death while walking across camp, of a bear sauntering into a barracks room and devouring a man as he slept.

Camps were the scenes of marriages, baptisms, first loves (and fourths); of dreams dreamed, dreams realized, dreams crushed. They were labeled in an Anchorage newspaper series as "The Skinny City." Bed availability in that skinny city peaked at 16,500. And anyone who was there knows there were times when few beds were empty.

Several changes were made to camp facilities during those three years. Many camps grew, both in acreage and capacity. Soon, those places, like the chunks of their lives spent there by workers, will be vague memories, some pleasant, some distasteful.

Of the 18 camps operated by Alyeska for construction of the pipeline, nine remain open. Six of those are scheduled for release either to Alyeska Surplus Management for disposal or to Operations Division by the end of the month, with two others to close during the first week of June, and the final camp to close by the end of June.

When actual construction on the project began in 1974 seven campsites had been used as staging areas. By July, four camps had been added and the original seven were expanded. **Five Mile,** with a 95-person capacity in January, by July was expanded to 260 beds. **Prospect Creek** Camp was expanded from 256 to 292; **Coldfoot** from 200 to 252; **Dietrich** from 196 to 296; **Galbraith Lake** from 264 to 312; **Toolik** from 240 to 340; and **Happy Valley** from 220 to 272. Those added during the first six months of 1974 were **Franklin Bluffs,** with an original capacity for 364; **Atigun** and **Chandalar** with 202 and 260 beds respectively, and **Old Man** with 463 beds.

In the fall of 1974 **Livengood** and **Delta** camps were constructed, each with a capacity for 1,120 persons; **Isabel Pass** Camp was built for 1,220; **Glennallen** for 672; **Tonsina** for 1,232, and **Sheep Creek** for 448.

As construction got under way in 1975, camps were "yo-yoed" to provide adequate bed space at the right spot at the right time, and it became difficult to keep track of camp sizes from one month to the next. Nearly half the barracks from Tonsina were removed, dropping capacity to 672; later, some barracks were returned. Glennallen was expanded to 1,300 beds; and late in 1975 a new camp was added. **Sourdough** was built with a capacity for 280 persons in December 1975,

(See Page 4, Col. 1)

PRUDHOE BAY
ARCTIC OCEAN
SLOPE
FRANKLIN BLUFFS
HAPPY VALLEY
TOOLIK
GALBRAITH
ATIGUN
CHANDALAR
DIETRICH
COLDFOOT
PROSPECT
OLD MAN
FIVE-MILE
LIVENGOOD
YUKON RIVER
FAIRBANKS
WAINWRIGHT
DELTA
ISABEL PASS
SOURDOUGH
GLENNALLEN
TONSINA
SHEEP CREEK
ANCHORAGE
VALDEZ
GULF OF ALASKA
CANADA
UNITED STATES

Galbraith Lake Camp In The Brooks Range

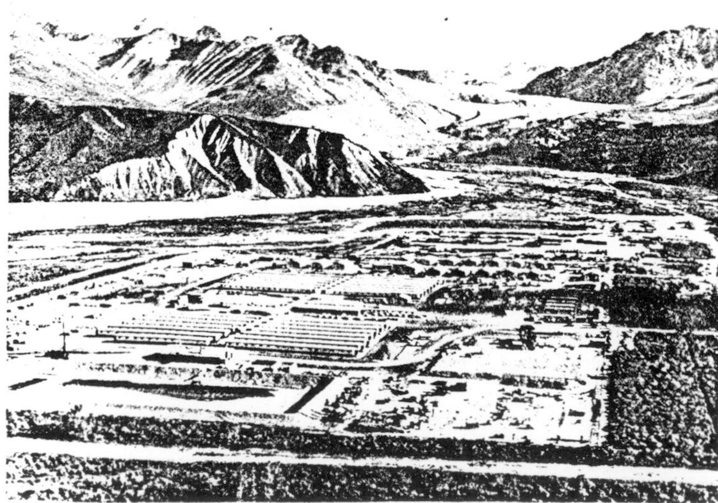

Isabel Pass Camp And Gulkana Glacier

Items Must Be Claimed By June

Pipeline construction personnel who have stored personal belongings at Alyeska Central Warehouse, Bay 15, at Fairbanks should contact warehouse personnel immediately, and must arrange to have all gear removed by June 15. According to Ian Herbert, manager of warehouse operations, the warehouse now contains personal effects of 80 persons for whom the warehouse employees have been unable to ascertain a current address.

Herbert explains that as construction nears an end, the existing storage space arrangements are being phased out. All items must be removed prior to June 15. Thirty days after that date any items which remain will be sold at a public auction and the proceeds will be given to charitable organizations.

Persons who have items stored at the warehouse should contact Warehouse Operations, 479-7995 or 452-7411, or on extensions 3091 or 4729, immediately to confirm their current address and their intention to remove items from storage prior to the June deadline.

Alyeska Organization

Corporate Name—Alyeska Pipeline Service Company. Alyeska is an Aleut word meaning "mainland."

Owner Companies and percentages of pipeline ownership.

BP Pipelines (Alaska) Inc.	50.0108%
ARCO Transportation Alaska, Inc.	22.2950%
Exxon Pipeline Company	20.3378%
Mobil Alaska Pipeline Company	3.0845%
Amerada Hess Pipeline Corporation	1.5000%
Phillips Alaska Pipeline Corporation	1.4158%
Unocal Pipeline Company	1.3561%
	100.0000%

Recoverable reserves, at discovery (estimated)

- Sadlerochit—9.6 billion bbl.
- Kuparuk—2.2 billion bbl.
- Lisburne—300 million bbl.
- Milne Point—100 million + bbl.
- Endicott—350 million bbl.
- Point McIntyre—300 million bbl.

Travel time at 1.46 million bbl./day and miles and line fill between stations

From	Hours	Miles	Linefill (bbl)
PS 1-2	10.73	57.76	653,853
PS 2-3	8.64	46.51	526,493
PS 3-4	7.39	39.78	450,310
PS 4-5	24.27	130.69	1,479,411
PS 5-6	14.89	80.20	907,864
PS 6-7	10.99	59.18	669,918
PS 7-8	13.95	75.10	850,132
PS 8-9	11.04	59.47	673,200
PS 9-10	6.89	37.08	419,746
PS 10-11	18.60	100.17	1,133,924
PS 11-12	9.12	49.10	555,812
PS 12-Valdez	12.11	65.23	738,404
Total	148.62	800.27	9,059,057

Costs

Construction of pipeline, pump stations, Marine Terminal (total costs less financing)—$8 billion

Operation of pipeline (1993)—$541.8 million approx.

Cost of construction if built in 1994, estimated—$20-22 billion (based on actual costs of pipeline, pump stations, Marine Terminal, and a 25% factor for additional environmental and quality program requirements plus contingency).

Throughput

Total per year through 1997

1977	112,300,000 bbl.
1978	397,008,750 bbl.
1979	467,776,770 bbl.
1980	554,934,043 bbl.
1981	556,067,441 bbl.
1982	591,141,545 bbl.
1983	600,858,560 bbl.
1984	608,836,116 bbl.
1985	649,886,953 bbl.
1986	665,434,992 bbl.
1987	716,662,005 bbl.
1988	744,107,855 bbl.
1989	688,062,255 bbl.
1990	654,551,673 bbl.
1991	665,174,678 bbl.
1992	639,363,127 bbl.
1993	591,222,326 bbl.
1994	579,422,677 bbl.
1995	555,938,859 bbl.
1996	525,506,504 bbl.
1997	487,094,963 bbl.
Cumulative total	12,051,249,996 bbl.

Fuel Gas Line

Location—Generally parallels mainline crude oil pipeline, from Prudhoe Bay to PS 4.

Function—Carries natural gas from North Slope fields to fuel turbines at pump stations north of the Brooks Range. (Turbines at stations south of the Brooks Range are fueled by liquid turbine fuel.)

Dimensions—

- Diameter—10 in. (34 miles from PS1 to PS2); 8 in. (115 miles from PS2 to PS4)
- Length—149 miles

Gas Temperature—30F max. (receipt at PS1)

Pig launching/receiving facilities—PS1, MP 34, PS4

Pressure—

- Current operating—1,090 psi
- Design—1,335 psi

Compressors—Two 1,200 hp gas turbine compressors at SPS 1 boost gas pressure from approx. 600 psi to 1,100 psi.